MW01629854

PREDICTING
WEATHER EVENTS
WITH ASTROLOGY

About the Author

Kris Brandt Riske (Arizona) is the executive director and a professional member of the American Federation of Astrologers (AFA), the oldest US astrological organization, founded in 1938. She is also a member of the National Council for Geocosmic Research (NCGR). She has a master's degree in journalism and a certificate of achievement in weather forecasting from Penn State.

Kris has been a speaker at various astrological conferences and has written for several astrological publications. She currently writes the annual weather forecast for Llewellyn's *Moon Sign Book* and is the author of *Llewellyn's Complete Book of Astrology* and *Llewellyn's Complete Book of Predictive Astrology*. In addition to astrometeorology, she specializes in predictive astrology.

Kris is an avid NASCAR fan and enjoys gardening and reading.

PREDICTING WEATHER EVENTS WITH ASTROLOGY

KRIS BRANDT RISKE, MA

Llewellyn Publications
Woodbury, Minnesota

First Edition
First Printing, 2014

Cover art: iStockphoto.com/11991746/© gremlin
       iStockphoto.com/13258938/© patrick heagney
       iStockphoto.com/2115339/© Sasha Radosavljevic
       iStockphoto.com/227772120/© lg0r2h
       iStockphoto.com/15100244/© Vectorig
Cover design by Ellen Lawson
Editing by Andrea Neff

Llewellyn Publications is a registered trademark of Llewellyn Worldwide Ltd.

The Chart wheels were produced by the Solar Fire Gold ver. 8 program, licensed for use by Astrolabe at www.astrolabe.com.

Page for January 2011 of *Astro America's Daily Ephemeris, 2000–2020 at Midnight* is reprinted with the kind permission of David R. Roell at Astrology Classics (www.astroamerica.com).

Library of Congress Cataloging-in-Publication Data
Riske, Kris Brandt.
  Predicting weather events with astrology / by Kris Brandt Riske, MA. —
First Edition.
     pages   cm
  ISBN 978-0-7387-4158-1
1. Astrology and meteorology. 2. Astrometeorology. 3. Weather forecasting. I. Title.
  BF1729.M47R57 2014
  133.5'855163—dc23
                               2014026218

Llewellyn Worldwide Ltd. does not participate in, endorse, or have any authority or responsibility concerning private business transactions between our authors and the public.

All mail addressed to the author is forwarded, but the publisher cannot, unless specifically instructed by the author, give out an address or phone number.

Any Internet references contained in this work are current at publication time, but the publisher cannot guarantee that a specific location will continue to be maintained. Please refer to the publisher's website for links to authors' websites and other sources.

Llewellyn Publications
A Division of Llewellyn Worldwide Ltd.
2143 Wooddale Drive
Woodbury, MN 55125-2989
www.llewellyn.com

Printed in the United States of America

Other Books and Articles by Kris Brandt Riske, MA

Llewellyn's Complete Book of Predictive Astrology
(Llewellyn Publications, 2011)

Llewellyn's Complete Book of Astrology
(Llewellyn Publications, 2007)

Mapping Your Travels & Relocation (co-author)
(Llewellyn Publications, 2005)

Mapping Your Money
(Llewellyn Publications, 2005)

Mapping Your Future
(Llewellyn Publications, 2004)

Astrometeorology: Planetary Power in Weather Forecasting
(American Federation of Astrologers, 1997)

Llewellyn's Moon Sign Book (contributor)
(Llewellyn Publications, 2002–2015)

Llewellyn's Sun Sign Book (contributor)
(Llewellyn Publications, 2007–2013)

Llewellyn's Starview Almanac (contributor)
(Llewellyn Publications, 2005–2006)

Civilization Under Attack (contributor)
(Llewellyn Publications, 2001)

To Nancy McEwen

Contents

CHARTS

Introduction

Nothing can be as pleasant or as devastating, as serene or as severe, as predictable or as extreme, as the weather. To a great extent, weather dominates and controls our everyday lives, whether it be a blizzard, a flood, or a simple rain shower that interferes with outdoor activities. Weather conditions affect transportation and communication and are critical to farming.

Research in atmospheric science and the development and refinement of tools such as radar and computer modeling have increased the ability of meteorologists to accurately forecast weather. Radar can now pinpoint the development of a tornado within about twenty minutes, warning people who might be in its path. And tornado prediction begins even before that, with the highly skilled meteorologists at the Storm Prediction Center in Norman, Oklahoma, who use satellite imagery and computer modeling to help forecast the potential for severe weather as many as several days in advance.

However, even with knowledge, skills, and available tools, the outside limit of an accurate meteorology forecast is only about five days, and usually less than that. Seasonal forecasts are becoming more dependable but do not approach the accuracy of mesoscale forecasting (short-term events such as thunderstorms, squall lines, fronts, and tornadoes that range in size from about a mile to 600 miles in size). The same is true of hurricane forecasting, which has not reached the accuracy level of mesoscale forecasting, particularly in tracking the path of a hurricane in order to determine where it will make landfall. The reason for this might be that because mesoscale events are more frequent, more funding has been invested in research and development of forecasting tools.

Enter astrometeorology, the art of weather prediction using astrology.

Astrometeorology has no time restrictions. Weather can be forecast for tomorrow, next week, next month, next year, or a date far in the future. Like meteorology, however, astrometeorology has its limitations. It is at its best when used to forecast weekly and seasonal weather trends and major weather events. It can be used to forecast a high-risk tornado day, week, or season, for example, but generally not the exact minute and location where a tornado will occur. Not that it can't be done; rather, few people have the time required to pinpoint the exact time and location. And why would we? Radar delivers a high level of accuracy. When the best of meteorology and astrometeorology are combined, the forecasting possibilities are limitless.

In addition to the potential to save lives, long-range weather forecasting can help cities, states, and even countries plan ahead for major weather events. Farmers can manage crops according to seasonal weather trends, and utility companies can do the same with energy demand. Investors, particularly those involved in the commodities market, can see increased gains, and retailers can determine whether to stock more umbrellas or snow shovels.

Astrometeorology can be used months and years in advance to identify the areas—regions, states, and locations within states—that will be prone to severe weather. For example, astrometeorology can be used to forecast the specific areas most prone to hurricane activity during a given year, such as the Florida Panhandle rather than just the East Coast, or coastal Louisiana and Mississippi rather than just the Gulf Coast. Specific areas especially prone to tornadic activity can also be forecast, as can a hot summer or a cold winter, a drought or a flood, or a high number of cyclonic storms in the Midwest or New England.

How Astrometeorology Works

It's easy to identify and explain the planetary factors associated with various weather events. For example, Neptune indicates warmth, precipitation, and flood potential; Saturn is active in major storm systems; Uranus identifies high-pressure systems; and Mars relates to heat.

However, there is a critical concept to understand before delving into the "how" of astrometeorology. Planets do not *cause* the weather; they *reflect* weather conditions. In this way, astrometeorology is much like meteorology. Both are used to identify the cause and effect of specific atmospheric conditions that result in specific weather, and both are used to forecast the weather. The difference is that one is an art and one is a science. But that isn't quite true. Years of experience enable meteorologists to add the "art" component to forecasts. Astrologers forecasting the weather need to do the same—in reverse— by adding some science to the art.

It's possible to forecast the weather without any knowledge of meteorology, just as it's possible to forecast the weather without any knowledge of astrometeorology. Used in tandem, however, the two can result in a higher level of accuracy, especially in medium- and long-range weather forecasting. You'll thus learn some very basic meteorology in this book in order to better understand how the planetary influences relate to the weather.

Using astrology to create a weather forecast is fairly straightforward, but it can also be complex. A brief introductory explanation will suffice here, with greater detail to come in other chapters.

Forecasting is all about longitude and latitude (location); the seasonal solar ingresses that mark the start of spring, summer, autumn, and winter; and the monthly lunar phases (new Moon, full Moon, and two quarter Moons). When these three are considered in unison, a weather forecast can be generated for a specific day or week; the solar ingress indicates the seasonal trends. Each lunar phase is in effect for about a week, and each ingress for about three months.

A seasonal forecast for two locations for the winter of 1993 is a good way to illustrate how weather differs in different locations, depending on which planets are highlighted at a specific longitude and latitude. The planetary positions are identical by degree and sign, but their placement in the charts is different.

Look at the winter ingress chart for December 21, 1992, for Chattanooga, Tennessee (Introduction 1). The vertical axis, or meridian (Midheaven/IC), with 18 Scorpio/Taurus on the cusps, is the longitude of Chattanooga. The horizontal axis, or horizon (Ascendant/Descendant), with 28 Capricorn/Cancer on the cusps, is the latitude of Chattanooga.

In this chart, there is one planetary alignment that indicates a winter featuring major low-pressure, cyclonic storms with abundant precipitation: a Venus-Saturn conjunction at 14–15 Aquarius. What made this significant to Chattanooga's weather was that the conjunction formed a square with that city's longitude (Midheaven/IC) at the time of the ingress. Chattanooga received twenty inches of snow during the Superstorm of March 12–14, 1993, which brought stormy conditions to much of the eastern third of the United States.

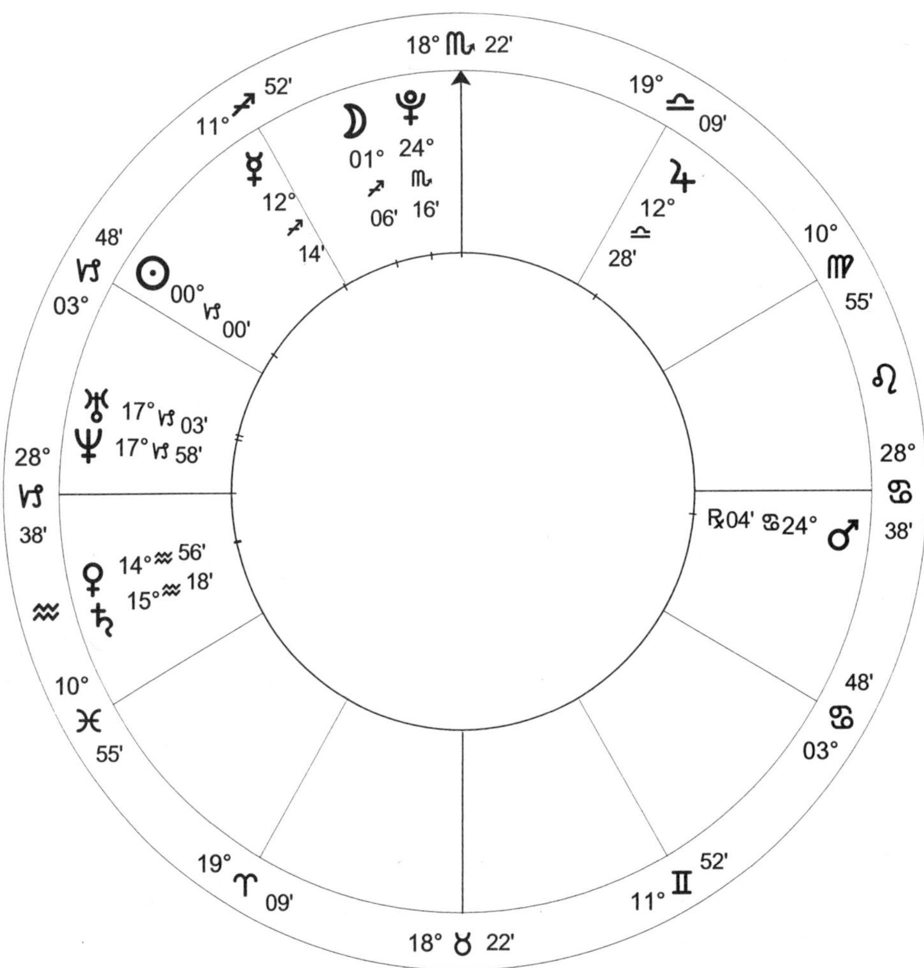

Introduction 1: Winter Ingress / Natal Chart / December 21, 1992, Mon / 9:43:14 am EST +5:00
Chattanooga, TN / 35°N02'44" 085°W18'35" / Geocentric / Tropical / Placidus / Mean Node

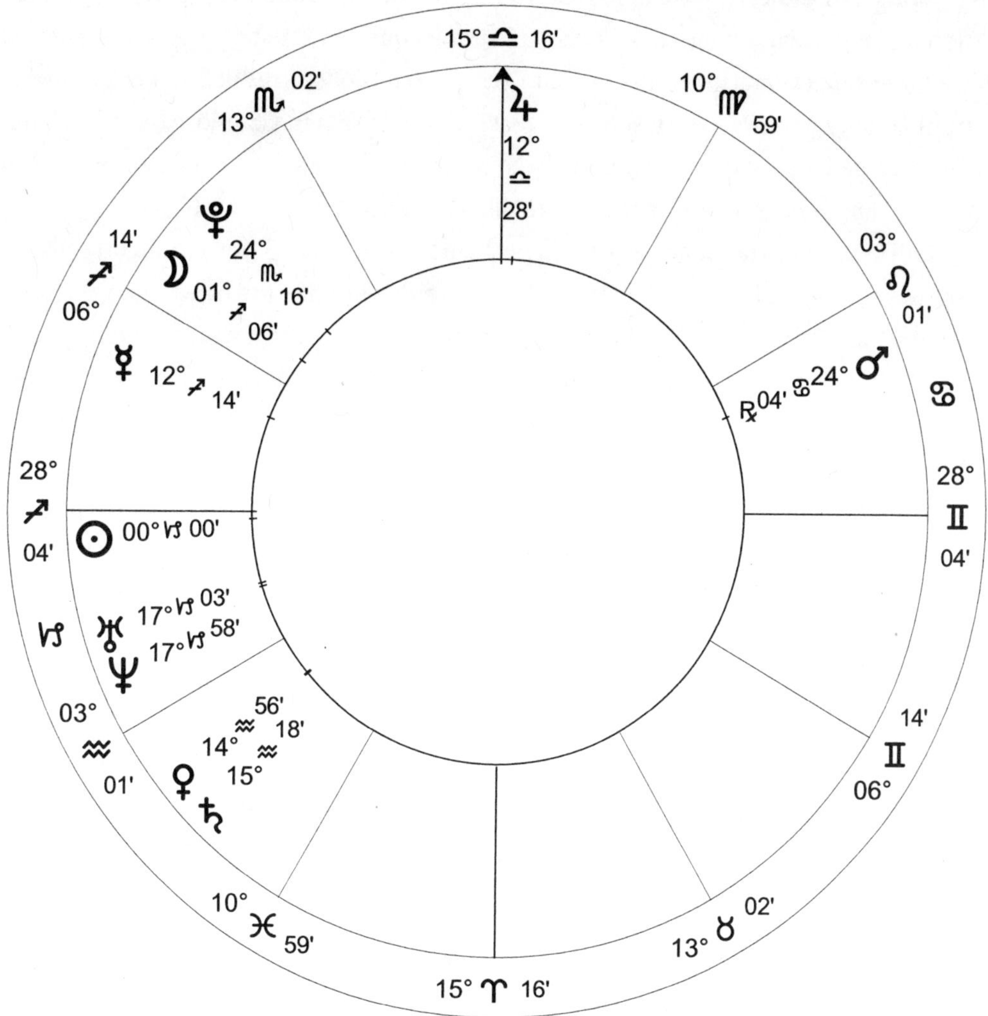

Introduction 2: Winter Ingress / Natal Chart / December 21, 1992, Mon / 6:43:14 am PST +8:00 San Diego, CA / 32°N42'55" 117°W09'23" / Geocentric / Tropical / Placidus / Mean Node

Now look at the same winter ingress chart for San Diego, California (Introduction 2), and notice that the planets are the same but the meridian (Midheaven/IC) and horizon (Ascendant/Descendant) are different by sign and degree. This is because San Diego is at a different longitude and latitude than Chattanooga.

This chart indicates a trend for cold, damp weather and abundant precipitation with flooding potential because the Uranus-Neptune conjunction at 17 Capricorn is square the ingress meridian (longitude) at San Diego. A series of storms in Southern California from January 5 to 19, 1993, resulted in 18.27 inches of rain in Campo, east of San Diego. The average January rainfall for San Diego is about two inches; in January 1993, the city received about nine inches of rain.

Obviously Chattanooga wasn't in a permanent storm condition throughout the entire winter, nor did San Diego see buckets of rain every day for three months. Nevertheless, the ingress charts reflected that winter's seasonal trends, which were indeed the norm and not the exception. An ingress chart shows the seasonal trend but not the timing, which is revealed in the lunar phase chart in effect at the time of a weather event. This is why San Diego experienced heavy precipitation in January, while Chattanooga saw its biggest storm in March.

This brief introduction to astrometeorology shows how planetary influences aligning with different longitudes and latitudes reflect varied weather conditions in each location. In later chapters we'll look more closely at a variety of weather events, along with the pertinent lunar phase charts for each.

There is, however, one other factor that is always wise to keep in mind when forecasting weather: local conditions. Rain is rare in southern Arizona in June, and temperatures in the 80s are unlikely in North Dakota in January. Familiarize yourself with the climatology of the forecast location, just as meteorologists do. With that knowledge and some practice, you can learn to forecast the weather for any location on any date.

CHAPTER 1

Weather Charts

A weather forecast requires two charts: one for the solar ingress and a second for the lunar phase. Both are calculated for the longitude, latitude, and time zone of the specific location for which you want to develop a weather forecast.

SOLAR INGRESS CHART

Each year has four solar ingresses, each of which marks the beginning of a new season. Spring begins when the Sun reaches 0 degrees Aries, summer with the Cancer ingress in June, autumn with the September Libra ingress, and winter with the Capricorn ingress in December. The ingresses occur around the 21st of the month. An ingress chart is in effect for the three months that follow, until the next ingress occurs, and is used to forecast the general weather conditions for the season at a specific location. For example, a Cancer ingress chart might indicate a cooler, wetter, warmer, or drier than normal summer.

In the 1902 winter ingress chart for Bismarck, North Dakota (Weather Charts 1), Neptune is prominently placed conjunct the IC (longitude). This planet of heavy precipitation and flooding opposes the Sun, Venus, and Mercury in the tenth house (longitude). The forecast would thus be for a wetter than normal winter for the area, with an increased number of storms with high winds, because Mars conjunct the Descendant (latitude) is square the Sun and Neptune.

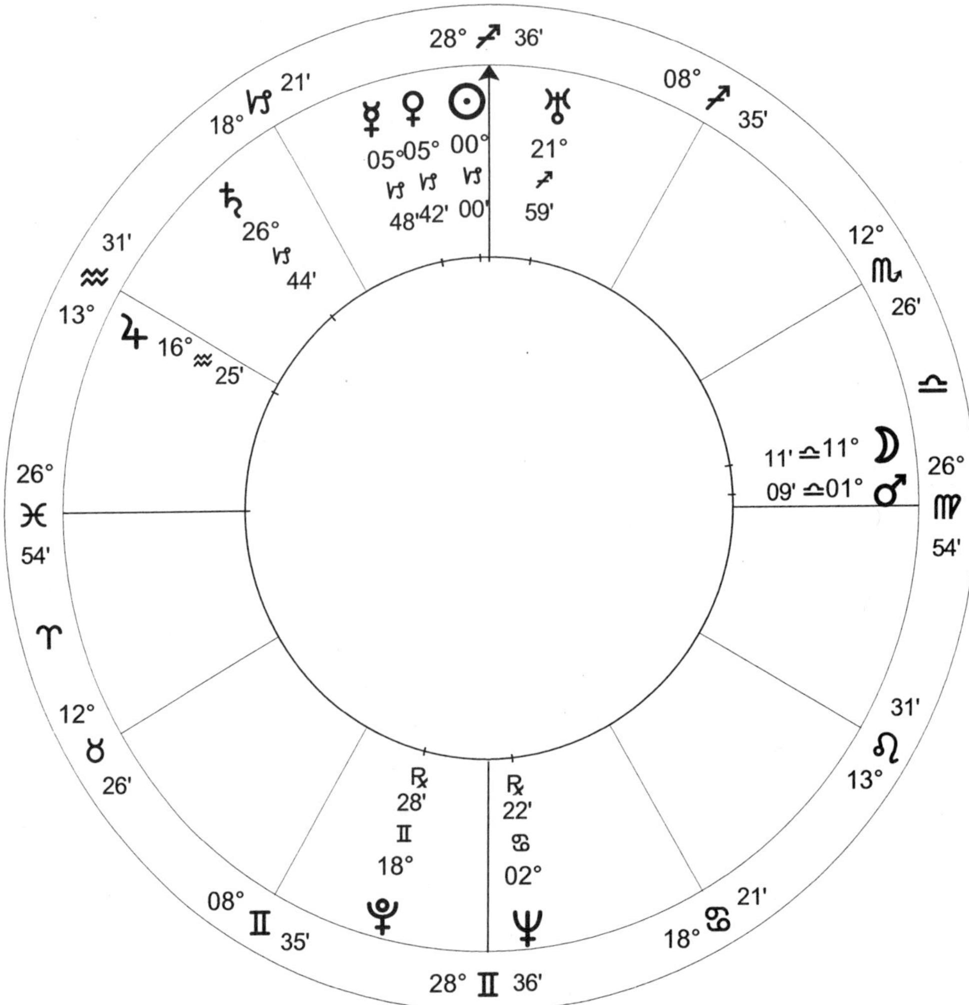

Weather Charts 1: Winter Ingress / Natal Chart / December 22, 1902, Mon / 12:35:32 pm CST
+6:00 Bismarck, ND / 46°N48'30" 100°W47' / Geocentric / Tropical / Placidus / Mean Node

An ingress chart calculated for the beginning of a specific season is used to forecast that season's weather trends. It also indicates any tendency for certain weather events. Although rain will usually occur at a location where dryness is indicated at the ingress, it will be below average and less frequent. A location that shows a winter tendency for major storms won't experience one every week during the three-month season, but overall the season is likely to be notable for snowfall totals.

The seasonal trends indicated in the ingress chart are important when forecasting weather for a specific day or week through the use of the lunar phase chart. For example, if an ingress chart forecasts temperatures well below normal, the weather indicated by a lunar phase chart that promises high heat will be warmer, but not excessively so. Rather, it might be experienced as a pleasant day in an otherwise cool summer.

The lunar phase chart is invaluable in forecasting the timing of major weather phenomena indicated in the ingress chart. For example, when planets in a lunar phase chart aspect several planets in that season's ingress chart, the lunar phase planets act as a trigger to activate the seasonal trend.

LUNAR PHASE CHART

The lunar phase chart is used to identify weather conditions during a more narrowly defined time period of about a week. There are four or five lunar phases each month, spaced approximately seven days apart. They occur at the new Moon, full Moon, and two quarter Moons. The new and full Moons are separated by approximately two weeks, with the quarter Moons occurring in between. Each phase is in effect until the next lunar phase, about one week later.

When forecasting, the new Moon has no more influence than the full Moon, and the same is true of the quarter Moons. All lunar phases are of equal strength and indicate the weather only during that phase. A lunar phase chart forecasts general conditions for the succeeding week in much the same way the ingress chart does for the season.

An example of a lunar phase chart is one for January 6, 1903, for Bismarck, North Dakota (Weather Charts 2), a location that had a seasonal trend for abundant precipitation that year. This chart forecasts precipitation because Neptune is conjunct the Ascendant (latitude). In the lunar phase chart, the Venus-Saturn conjunction (heavy precipitation) aspects the meridian (longitude) and the horizon (latitude) of the winter ingress chart. On January 6 and 7, a blizzard with 58 mph winds buried Bismarck in snow.

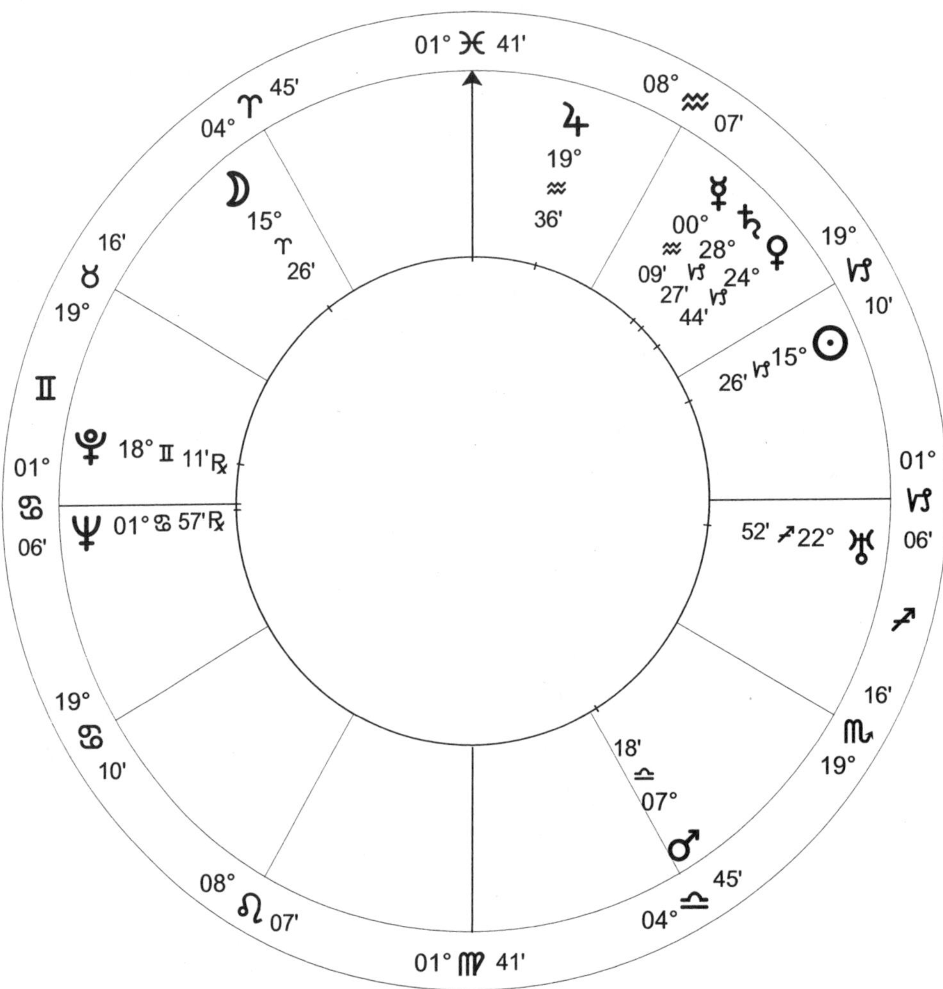

Weather Charts 2: Lunar Phase / Natal Chart / January 6, 1903, Tue / 3:56:44 pm CST +6:00
Bismarck, ND / 46°N48'30" 100°W47' / Geocentric / Tropical / Placidus / Mean Node

Ingress–Lunar Phase Comparison

Both the ingress and lunar phase charts must be used when forecasting. Of the two, the lunar phase chart is generally more powerful in the sense of immediacy, while the ingress chart sets the fundamental conditions of the season. However, the ingress chart is nearly always activated when a weather event occurs, even a minor one such as rain showers.

If a summer ingress chart indicates blazingly hot weather for the season, a lunar phase chart with Mercury conjunct the IC is unlikely to indicate much of a drop in temperature. However, if the same lunar phase chart was partnered with an ingress chart that also had Mercury conjunct the IC, it would mean temperatures well below the seasonable norm during that week, most likely from a strong cold front moving through the area.

In the Bismarck, North Dakota, example, the forecast for the week of January 6 indicated heavy precipitation because Neptune, already strongly placed in the ingress chart, was also strong in the lunar phase chart. Venus in the ingress chart was activated by a semisquare from Jupiter in the lunar phase chart; Jupiter, the planet of expansion, also indicated heavy downfall. The Venus-Saturn conjunction in the lunar phase chart was sextile the ingress chart's Ascendant and semisextile the Midheaven. When you see multiple interchart aspects such as these, it indicates the potential for a major weather event.

Transiting Aspects

The ingress and lunar phase charts are the foundation of weather forecasting, and a third component, the transiting planets, is used to narrow the timing of weather events. Often the transiting Moon is the trigger that times the beginning or ending of a weather event. Other fast-moving planets can indicate timing regarding the length of a major storm such as a blizzard or hurricane.

Transiting planets are considered on two levels: their aspects to each other, and their aspects to the ingress and lunar phase charts.

Transiting planets and their aspects to both charts and to each other have a significant influence both in terms of weather and the timing of a weather event. The inner planets—Sun, Moon, Venus, Mercury, and Mars—are usually the ones that trigger ingress and lunar phase chart aspects, marking the onset of weather.

The Moon in Aries was a key player in the Bismarck example, activating the lunar phase Venus-Saturn conjunction through a square. It was also square ingress Saturn. On January 7, transiting Venus formed a conjunction with ingress Saturn and transiting Mercury was trine ingress Mars, activating the ingress Mars-Neptune square, another aspect of heavy precipitation. Again, multiple aspects can alert you to a major weather event.

Local Conditions

Local conditions must always be considered when forecasting weather. Seasonal norms are readily available online from the National Weather Service (www.weather.gov) and at Weather Underground (www.wunderground.com), by far the best site for weather information, including local radar, satellite images, information regarding tropical storms, and educational material. The Weather Channel (www.weather.com) airs forecasts and covers major weather events. (See the resources section at the end of the book for more information.)

Longitude and Latitude

When forecasting weather events, the twelve astrological houses are not considered in the traditional manner, such as the third house giving indications for communication and the tenth house for career. Rather, the houses indicate direction—north, south, east, and west—from the longitude and latitude of the point of observation (location for the forecast).

Longitude is the meridian, the north/south axis traditionally referred to as the Midheaven/IC. Latitude is the horizon, the east/west axis known as the Ascendant/Descendant. Viewed from the center of the chart, the Midheaven (top of the chart) is south, and the IC (bottom of the chart) is north. This is the opposite of the way locations are viewed on a map, where the top represents north and the left represents west. The astrological chart is "flipped" because the Ascendant, which is the sign rising over the horizon, rises toward the Midheaven, just as the sun at dawn (represented by the Ascendant) rises to its high point at noon (represented by the Midheaven). A note on the Southern Hemisphere:

the directions are reversed. The Midheaven is north, the IC is south, the Ascendant is west, and the Descendant is east.

Longitude begins at 0 degrees at Greenwich, England, and moves around the globe to 180 degrees east longitude before it begins to descend from 180 to 0 degrees west longitude, returning to its starting point at Greenwich. This means that locations from the North Pole to the South Pole, whether north or south of the equator, have the same longitude. Thus, if a weather chart for Chicago has a 15 Scorpio Midheaven, so does New Orleans, which is directly south, and Green Bay, Wisconsin, which is directly north. The Midheaven degrees are the same for all locations at that longitude, whether they are north or south of the equator. Longitude is measured in degrees east or west of Greenwich; the International Date Line is 180 degrees longitude.

However, the degrees on the east/west axis change as the latitude moves north or south of a point of observation. Viewed from the center of the chart, the Ascendant is the eastern horizon and the Descendant is the western horizon. Latitude is measured in degrees north or south of the equator, with the equator being 0 degrees and the North and South Poles being 180 degrees. Locations that are the same number of degrees south of the equator have the same signs on the Ascendant/Descendant axis, but in reverse; thus, if Virgo is on the Ascendant in the Northern Hemisphere, Pisces is on the Ascendant in the Southern Hemisphere at the same number of degrees south of the equator.

Using the same Chicago example, suppose the Ascendant (latitude) is 15 degrees Aquarius. But New Orleans, farther to the south, has an 18 degree Aquarius Ascendant, while Green Bay's Ascendant is 13 degrees Aquarius. The degrees increase to the south, because when the Ascendant reaches 30 degrees Aquarius, it will move to the next or higher sign in the natural order of the zodiac, which is Pisces, starting over at 0 degrees.

ANGLES AND DIRECTIONS

The Midheaven, IC, Ascendant, and Descendant are the chart angles. It is these points that are of utmost importance in astrometeorology because they are true only for a specific location either north or south of the equator and either east or west of Greenwich.

The two points of the Midheaven/IC, or north/south (vertical) axis, are two of the angles in a weather chart. Points to the east are represented by the Ascendant, while those to the west are found at the Descendant. The houses in between represent these directions: southwest is the eighth house, northeast is the second house, southeast is the eleventh house, etc. However, it is most often the horizon and/or meridian that most accurately reflects the weather at the point of observation.

A weather chart with an active meridian (many planets in the fourth or tenth house or conjunct the IC from the third house or conjunct the Midheaven from the ninth house) is more likely to indicate active weather conditions at the designated location. This is even more true when those planets form opposition aspects to planets in the ninth and tenth houses or when planets in the first and seventh houses are square the meridian.

Picture a chart for Philadelphia (75 degrees west of Greenwich) that has 15 Taurus on the IC, with Saturn located in the third house at 15 Aries. The third house indicates an area west of Philadelphia, so when the point of observation is moved 30 degrees west (the chart is turned counterclockwise) to Denver, both Saturn and the IC are at 15 degrees Aries. With this simple calculation, you would know that Denver will experience cloudy skies and temperatures below normal (Saturn influence).

While this example may appear to be contradictory, it is not. From direction alone, it seems that the weather would occur to the east of Philadelphia because the third house is in the eastern half of the chart. However, when the chart wheel is rotated counterclockwise, the sign on the IC moves to the right, toward the Descendant.

Since this is true, when planets are in the third house (the eastern half) of the chart, the weather represented by those planets is west of the point of observation but moving east toward the location's longitude. And if the planets are located in the ninth house (on the western side of the chart), they are also moving to the east. In other words, when the chart is turned counterclockwise, planets in the third house move east (toward the Descendant, the eastern point of the chart) until they reach the IC. And that is exactly how the majority of weather systems move in the Northern Hemisphere: west to east.

The following chart illustrates this principle (Weather Charts 3). When the wheel is rotated 15 degrees (counterclockwise), the meridian axis changes from 0 Pisces to 15 Aquarius. Therefore, at longitude 90W (St. Louis, Missouri), Uranus will conjunct the IC, with the Sun conjunct the Midheaven opposing it. It is along this axis that the effects of Sun opposition Uranus (cooler, cloudy, precipitation) will have maximum impact. Rotating the wheel farther west to 120 degrees (West Coast) puts Mars on the axis, the point at which its effects will be observed. It will be cool in St. Louis, but hot in California.

The third/ninth-house opposition of the Sun and Uranus also indicates weather that is approaching the 75 degree axis (East Coast) from the west. The weather pattern is currently to the west of the point of observation and moving toward it. Similarly, where the Midheaven/IC is 0 degrees Leo/Aquarius (105W), hot weather (Mars) is approaching from the west.

This is how you can determine where the maximum effects of weather indicators will occur. It is also why, for example, it can be snowing in Kansas City, Missouri, while mild conditions prevail in Boston, Massachusetts. This concept will be further explored in other chapters, particularly chapter 3 on national and local weather forecasting.

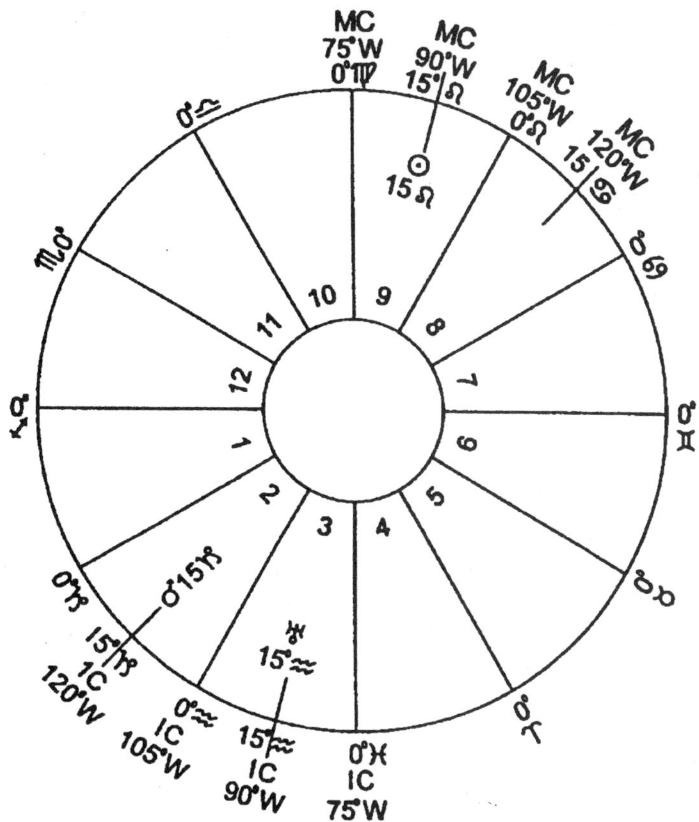

Weather Charts 3: Latitude and Longitude Example

CHAPTER 2

Planets, Aspects, and Signs

Weather is identified primarily by the planets, their aspects to each other, and their placement in the chart. Planets that are within about 4 degrees of the chart angles at a specific location or those that are square the meridian or horizon at a specific location are the most powerful. The tighter the orb, the stronger the aspect. Each planet has a particular function in weather forecasting, but they all take on additional characteristics when in contact with other planets.

In addition, you should look at the planets to the left of the IC and the right of the Midheaven because they often offer additional information that can be pertinent to a forecast. Planets to the left of the IC and the right of the Midheaven indicate weather that is approaching from the west, while planets to the right of the IC and to the left of the Midheaven reflect weather to the east of a location. So if you see Mercury about 7 degrees to the left of the IC in a lunar phase chart, you could forecast windy conditions later in the week when Mercury crosses the IC.

SIGNS

The signs of the zodiac play a less important role in astrometeorology than they do in other branches of astrology. However, there are some exceptions to that generality, such as the influence of the water signs (Cancer, Scorpio, Pisces) and their relationship to precipitation, the barren signs (Aries, Gemini, Aquarius, Libra, Virgo, Sagittarius) as they relate to lower levels of precipitation, and the inclination of Aries and Leo to higher temperatures. Although the signs generally have only a secondary effect, at times they can have a compounding one, such as Neptune (precipitation) in a water sign or Mars (heat) in a fire sign. In addition, a planet in the sign it rules tends to amplify the meaning of that sign and planet; for example, Mars in Aries emphasizes the heat factor, while Saturn in Capricorn indicates temperatures that can be far below normal.

- **Aries**, ruled by Mars: dry, warm/hot, wind
- **Taurus**, ruled by Venus: precipitation, moderate temperatures
- **Gemini**, ruled by Mercury: dry, cool, wind
- **Cancer**, ruled by the Moon: cold, wet
- **Leo**, ruled by the Sun: dry, hot
- **Virgo**, ruled by Mercury: cold, dry, blustery
- **Libra**, ruled by Venus: dry, cool, breezy, sunny, fair
- **Scorpio**, ruled by Pluto: cloudy, cool/cold, wet
- **Sagittarius**, ruled by Jupiter: dry, fair, warm
- **Capricorn**, ruled by Saturn: cold, stormy, wet
- **Aquarius**, ruled by Uranus: cold, dry, high pressure (high or low temperatures)
- **Pisces**, ruled by Neptune: wet, cool, low pressure

You can draw some correlations by considering the signs from an elemental perspective:

The **fire signs**—Aries, Leo, and Sagittarius—tend toward warmer, drier weather.

The **earth signs**—Taurus, Virgo, and Capricorn—tend toward cooler weather with precipitation.

The **air signs**—Gemini, Libra, and Aquarius—tend toward windy weather that can be warm or cool.

The **water signs**—Cancer, Scorpio, and Pisces—tend toward cooler, wetter weather.

In addition to looking at the sign of a pertinent planet in an ingress or lunar phase chart, the signs on the Midheaven/IC and Ascendant/Descendant axes can provide information about general weather conditions in a specific location. This is particularly true of the ingress chart, which is used in part to forecast seasonal trends regarding temperature and precipitation levels.

PLANETS

Sun

The Sun governs the constitution of the weather and basically has no qualities of its own. In a way, it functions as a trigger, because when it aspects planets other than the Moon, it generally indicates the kind of weather associated with the other planet. For example, the Sun with Saturn indicates stormy conditions. There is an exception to this, however. The Sun can indicate higher temperatures when it's conjunct the IC or Midheaven and does not contact a planet of cool or wet weather.

Moon

Because the Moon moves so quickly, it functions as a trigger, setting off other aspects. If, for example, Venus in Cancer is square the meridian in a lunar phase chart, precipitation will often begin when the transiting Moon aspects Venus. The Moon has no inherent qualities of its own, but if it's conjunct the lunar phase IC in a sign associated with precipitation, then rain or snow is more likely to occur during that week.

Mercury

Mercury can also serve as a trigger, activating planetary aspects, because of the speed with which it moves. Primarily concerned with wind direction and velocity, Mercury is often active where weather fronts occur. This is particularly true of the locations where Mercury aspects the horizon at the ingress; many storm fronts will travel along that horizon line (generally west to east) during the season. Mercury's effect is modified by the planets it aspects, and the planets it forms aspects with determine direction and velocity.

When conjunct the IC in an ingress or lunar phase chart, Mercury indicates unseasonably cool weather and strong winds of varying frequency. This planet is also generally active where strong thunderstorms, tornadoes, hurricanes, and cyclones occur.

(Note: wind direction is always identified as the direction from which the wind is blowing; for example, a northerly wind is one that blows from the north to the south.)

Venus

Venus's general tendency is for moisture and, to some extent, clouds. It's also a planet of humidity, warmer temperatures, and pleasant, sunny weather. If Venus is conjunct the Midheaven in an ingress or lunar phase chart, storms will be centered to the south (Midheaven is south). When Venus is conjunct the IC, precipitation will be centered to the north, and the south will experience rising temperatures. This is especially true in winter months, when it can indicate heavy snow or freezing rain.

Nevertheless, a planet conjunct the IC or Midheaven can indicate weather at the forecast location, so it's wise to strongly consider its influence when it's placed at either of the longitudinal angles. Venus conjunct the IC can thus indicate precipitation to the north or south, just as it can when conjunct the Midheaven.

Venus is quite cold when in Aquarius and is drier in the other barren signs (Aries, Libra, Virgo, Gemini, Sagittarius). When placed in any of the water signs (Cancer, Scorpio, Pisces) or in Taurus, precipitation can be excessive. Venus (or Neptune) aspecting the longitude or latitude in the center of Tornado Alley at the spring ingress indicates a season with a strong northward flow of warm, moist air from the Gulf, a necessary thunderstorm component.

Mars

Rising temperatures, westerly winds, and evaporation are characteristic of Mars. This planet is nearly always active during intense heat waves and drought, especially when in Leo. When Mars is at maximum perigee (its closest approach to the Earth) every fifteen years, summers are abnormally hot in areas where it aspects the meridian (1971, 1986, 2001, 2016, etc.). When configured with colder planets (Mercury and Uranus), Mars reflects the heat and energy necessary to generate a storm.

Mars's action is intensified to the north when it occupies the lower meridian in an ingress or lunar phase chart and to the south when it is conjunct the IC, but as with other planets, this is not always true. Mars conjunct the Midheaven can just as easily indicate hot, windy weather to the north, depending on other aspects. So, again, carefully study aspects and contacts with the ingress or lunar phase chart. Humidity sometimes rises when Mars is in one of the water signs or aspects Venus or Neptune. Mars is also associated with windy conditions and is often active where thunderstorms, hurricanes, and tornadoes occur.

Jupiter

Jupiter indicates air descending from the upper atmosphere and high-pressure systems, making it generally a fair weather planet. When in northern declination (see the following section, "Aspects and Orbs") in Aries or Leo, Jupiter is associated with higher temperatures. In other configurations, it produces cold weather and high winds. This is because Jupiter amplifies and expands the influence of the planet it aspects. On its own, and conjunct or square the meridian or horizon, it indicates elevated temperatures. When configured with Mars or Pluto, Jupiter can be associated with drier conditions or drought, but also with powerful storms. Jupiter is often active in thunderstorms because it's associated with descending, cooler air, which thunderstorms require.

Saturn

Unlike the quick action of Mars, Saturn is associated with slow-moving weather events and systems, concentrating its effects in a wider area or region. It is thus indicative of more prolonged and widespread storms. Saturn is a low-pressure planet, and its cold, cloudy, wet nature is centered where it is conjunct or square the meridian. When Saturn is conjunct the IC, storms tend to shift northward while temperatures rise in the south and southeast. When it is conjunct the Midheaven, storms are centered to the south as cold air from the north, northwest, or west moves into southern areas.

Saturn skies vary from partly cloudy to overcast, and this planet is associated with the characteristic counterclockwise air flow of low-pressure systems. When Saturn is in aspect with Mercury or Venus in Capricorn or Aquarius, much colder weather can be expected. Saturn is often active in cyclones and hurricanes.

Uranus

Uranus indicates high pressure and descending cold air from the Arctic and the upper levels of the atmosphere. Always indicative of colder temperatures and cold waves in cold months, Uranus is often active in storms, cyclones, hurricanes, strong thunderstorms, and tornadoes, and can indicate cold, gusty winds. Uranus conjunct the IC in an ingress or lunar phase chart indicates unseasonably cold temperatures as well as below-normal precipitation and isolated droughts (some areas experience above-normal precipitation because of cold fronts).

When Uranus is conjunct the Midheaven, the cold, dry air will extend far south, and low-pressure systems will be to the north. If Uranus is square the meridian, high- and low-pressure systems converge, developing storms and wind. In warmer seasons, particularly summer, Uranus is prominent—aspecting the angles—during periods of dry, hot weather, with cooler, wetter weather occurring to the east or west. Sometimes this is referred to as a "blocking high," because the high-pressure system is "parked" in a large region, preventing moist and cooler air from entering the area. Generally, Uranus represents colder temperatures in colder seasons and warmer temperatures in warmer seasons.

Neptune

Neptune indicates counterclockwise, ascending air currents and southerly air flow, humidity, and milder temperatures. Also a planet of low pressure, it is often active in foggy conditions and usually has a presence in hurricanes and tornadoes. Depending on the configuration, Neptune, the ultimate planet of moisture, can indicate squalls, thunderstorms, hurricanes, tornadoes, and excessive precipitation. Severe flooding and overcast skies often occur where Neptune strongly aspects the meridian of an ingress or lunar phase chart. Neptune is also associated with winter and spring thaws. At times, Neptune is nebulous, with some areas receiving heavy precipitation and other areas very little. Years when Saturn and Neptune are in a major hard aspect (see the following section, "Aspects and Orbs") often produce precipitation well above normal and extreme flooding.

Pluto

Pluto increases the intensity of any configuration in which it is involved, and years with a strong Pluto aspect at an ingress are often characterized as extreme from a weather perspective. Pluto's effect regarding wind direction and velocity is quite prevalent in tornadoes, hurricanes, and cyclones. This planet is also indicative of both higher and lower temperatures, intense heat and cold. When Pluto is in strong aspect with Jupiter, for example, temperatures can be blistering, well outside the norm, and drought is often a major concern.

ASPECTS AND ORBS

In the context of astrometeorology, aspects can be divided into three basic categories—easy, hard, and neutral—although there are some exceptions. All aspects in each category operate similarly, the primary difference being their strength.

Aspects vary in strength, so a sextile is similar to but stronger than a trine, and a square is similar to but stronger than a semisquare. Strength in a weather chart can be defined as the weather indicated by the aspect a planet makes to the meridian or horizon: a brief shower versus a downpour, a 5 mph wind versus a 30 mph wind, or even a tropical disturbance versus a hurricane. So when you see Saturn square the meridian, you can forecast a stronger storm than when Saturn is semisquare/sesquisquare the meridian.

In order of strength, with the strongest first, the easy aspects in astrometeorology are the sextile, trine, semisextile, and inconjunct. Hard aspects are the opposition, square, semisquare, and sesquisquare. Neutral aspects are the conjunction and the parallel and contraparallel of declination. These aspects are classified as neutral because the effect depends on the planets in aspect. When planets are in both longitudinal and declination aspect, their strength multiplies. If you see, for example, Venus conjunct and parallel Saturn, storm precipitation levels are likely to be higher than if the planets were only conjunct.

When viewing easy aspects, you'll need to shift your thinking somewhat, as they don't operate exactly like easy aspects in a natal chart. In general, easy aspects indicate fair weather, hard aspects indicate the reverse, and neutral aspects can be either. With weather, though, skies can be fair under a dominant sextile or trine aspect, but also very cold. Some of the coldest winter weather is associated with the sextile and sometimes with the trine. Temperatures are low, yet skies are fair, a combination that many people would not classify as "good weather."

The same can apply to hard aspects. A Sun-Mars sesquisquare in summer can indicate warmer temperatures and dryness rather than stormy conditions. That's good weather if you like heat and are planning a trip to the lake, but it's bad weather if your garden plants are shriveling.

Important point: Generally, if the planets involved in an aspect do not aspect the horizon or meridian at the forecast location, the weather phenomenon they represent will not occur at that location; the weather will occur elsewhere, which could be anywhere in the world.

The use of the parallel and contraparallel of declination are valuable in astrometeorology. You can do a weather forecast without them, but if they aren't used, the intensity of weather phenomena can be easily missed. Declination can be especially useful in weather forecasting when planets are on the celestial equator (0 degrees of declination) or in maximum declination (20 degrees or more north or south of the celestial equator). In addition, the parallel or contraparallel intensifies longitudinal aspects more than twofold, especially when the planets are also in conjunction or opposition. Most astrological calculation software can generate an aspectarian that includes the declinations. Ephemerides produced by Astro America and the Rosicrucian Fellowship are the only ones that include planetary declination.

A 4-degree orb is allowed for longitudinal aspects to the meridian and horizon and between planets in the ingress and lunar phase charts. A 1-degree orb is used for declinations.

Planets near the horizon or meridian but not necessarily within orb can reveal entire regions affected by weather phenomena or indicate approaching weather. Approaching weather is indicated by planets to the left of the IC; if you mentally reverse the chart, viewing the Midheaven as south, then approaching weather is indicated by planets to the right of the Midheaven. Venus could, for example, be 6 degrees to the left of the IC in a lunar phase chart and transit during the week to form a conjunction with the IC; this could then be used to time a rain forecast.

A 1-degree orb is used for transiting planetary aspects. Usually these are the faster-moving planets (Sun, Moon, Mercury, Venus, Mars), which generally act only as they approach exactitude in their aspects to the meridian or horizon in an ingress or lunar phase chart. The time frames used in astrometeorology—three months or one week—are too short to result in much movement of the outer planets (Jupiter, Saturn, Uranus, Neptune, Pluto). The outer-planet configurations are longer term, and the weather phenomena associated with them are usually triggered by the faster-moving planets. However, when an outer planet aspects an ingress or lunar phase chart planet or angle within 1 degree, the weather effect is often more intense.

RETROGRADE, STATIONARY, AND ZERO OR HIGH DECLINATION PLANETS

Because the outer planets (Jupiter, Saturn, Uranus, Neptune, Pluto) are retrograde during as much as half their orbit, their retrograde periods have a minimal effect in weather forecasting. However, these planets have a very strong effect when they are stationing to change direction, either retrograde or direct.

The opposite is true of Mercury, Venus, and Mars, which take on extremely negative characteristics during their retrograde periods. When these three planets are retrograde, even easy aspects to or from them should be interpreted in the negative. Mercury retrograde can be associated with storm-related events such as power and communication outages, and a retrograde Venus can indicate increased precipitation. Retrograde Mars can accompany elevated temperatures, dryness, or storms.

The effects of all stationary planets (whether stationary direct or stationary retrograde) are generally negative, especially when they are in aspect to the horizon and meridian.

Planets at 0 degrees (low declination) or far north or south (20 degrees or more north or south in high declination) declination also bring those planets into high negative focus. When seven or more planets are in northern declination, temperatures in the Northern Hemisphere are higher; when seven or more are in southern declination, temperatures are lower overall in the Northern Hemisphere. The reverse is true for the Southern Hemisphere. Signs of northern declination are Aries, Taurus, Gemini, Cancer, Leo, and Virgo; southern declination signs are Libra, Scorpio, Sagittarius, Capricorn, Aquarius, and Pisces.

Individual planets are affected by these previously mentioned conditions in these ways:

Mercury's tendency toward wind is increased with precipitation and cooler temperatures in autumn and winter.

Excessive humidity and precipitation are characteristic of Venus, with cooler temperatures in autumn and winter.

Mars indicates atmospheric disturbance in summer and winter, heat waves in summer, and generally dry air in spring and autumn.

Jupiter is associated with elevated temperatures and amplifies the nature of other planets it aspects.

Saturn indicates overcast skies, excessive humidity, cooler temperatures, and persistent precipitation.

Weather phenomena associated with Uranus include cold fronts, colder temperatures (severe cold in winter), frost, storms, cloudy skies, and precipitation.

Although Neptune is still associated with warmer temperatures, humidity, and precipitation, these periods are ripe for unusual weather occurrences.

Pluto intensifies heat/cold, high winds, and storms.

ECLIPSES

Eclipses occur between two and seven times each year, but most years have four, sometimes five, eclipses. Eclipse dates are listed in ephemerides and on various Internet sites and can be found using astrological calculation software.

When you're studying an ingress chart, always look for the ingress chart longitudes and latitudes that receive a hard aspect from an eclipse, as these locations are often the site of a major weather event during the succeeding six to twelve months. But major weather events can occur a month or two prior to an eclipse that aspects an ingress chart, so also look for any upcoming eclipses within about six months after the ingress. Also check for eclipses that aspect the longitude and latitude in lunar phase charts, looking for eclipse aspects to these charts.

Sometimes one eclipse will be at the sign and degree of the meridian or horizon of an ingress chart and another eclipse will fall on the meridian or horizon of a lunar phase chart for the same location, or vice versa. At other times, only one chart is aspected by an eclipse. Those ingress and lunar phase chart angles that are aspected by one or more eclipses often indicate the location of powerful weather events.

For example, if Saturn is square the meridian in an ingress chart and an eclipse aspects the same meridian, storms in that location have the potential to be much stronger. The storm would be even stronger if another eclipse aspected the horizon in a lunar phase at the same location, and especially so if Saturn or other storm indicators were active in the lunar phase chart.

PLANETARY ASPECTS

Use the planetary configurations listed here as guidelines rather than representative of finite weather phenomena. Much depends on the interaction between the ingress and lunar phase charts and the transits to both, as well as the climatology of the location. Remember, too, for example, that if an ingress chart shows heavy seasonal precipitation, even an easy planetary aspect aligned with ingress Neptune conjunct the IC could indicate rain.

Again, the planetary aspect must aspect the longitude and/or latitude (almost always both if there is a strong weather event) of the location in order to reflect a weather event at that location.

Sun-Mercury

Easy Aspects: fair in all seasons; breezy in summer and windy in all other seasons; cooler in spring, summer, and winter.

Hard and Neutral Aspects: precipitation in all seasons; high wind in spring (gales and strong fronts with Mercury retrograde) and winter (blizzards with Mercury retrograde), breezy in summer, windy in autumn.

Sun-Venus

Easy Aspects: fair with rising or mild temperatures all seasons.

Hard and Neutral Aspects: warm, humid, and showers in spring and summer; cloudy, rainy, and cooler in autumn; foggy, abundant precipitation, and cooler in winter.

Sun-Mars

Easy Aspects: warmer in spring and autumn, hot in summer, moderate temperature rise in winter; evaporation in spring and summer, dry in autumn.

Hard and Neutral Aspects: warmer, storms all seasons.

Sun-Jupiter

Easy Aspects: northerly air flow all seasons; warmer and pleasant all seasons.

Hard Aspects: mild to warm all seasons; breezy in autumn and winter; thunderstorms in summer.

Neutral Aspects: warmer temperatures.

Sun-Saturn

Easy Aspects and Parallel-Contraparallel: partly cloudy in spring and autumn; frost in spring and autumn; breezy with showers in summer; fair in winter; colder all seasons.

Hard Aspects and Conjunction: low pressure, colder, cloudy, east to northeasterly winds, heavy precipitation and storms all seasons.

Sun-Uranus

Easy Aspects: cooler (sudden cold waves possible), fair, and northwesterly winds all seasons.

Hard Aspects: cooler, precipitation and storms, overcast, damp, windy, sudden changes all seasons.

Neutral Aspects: sudden change to cooler, clear; northwesterly winds all seasons.

Sun-Neptune

Easy Aspects: fair, warmer, calm to breezy.

Hard Aspects: fog, warm/hot, misty, humid, squalls, storms, precipitation, variable wind.

Neutral Aspects: warmer, unusual weather with sudden changes, ascending air flow, misty, fog.

Sun-Pluto

Easy Aspects: warmer all seasons.

Hard and Neutral Aspects: warmer or colder all seasons; storms all seasons.

Mercury-Venus

Easy, Hard, and Neutral Aspects: gentle breezes (usually southerly), warmer, moderate humidity, partly cloudy all seasons.

Mercury-Mars

Easy Aspects: warm and windy all seasons.

Hard and Neutral Aspects: windy with gusts and sudden changes all seasons; thunderstorms with hail potential in warm seasons, or hot and dry; seasonal temperatures in winter.

Mercury-Jupiter

Easy, Hard, and Neutral Aspects: warmer, fair, moderate northerly winds, cumulus clouds all seasons.

Mercury-Saturn

Easy Aspects: partly cloudy, cooler all seasons.

Hard and Neutral Aspects: increasing cloudiness, low pressure, counterclockwise air flow, humidity or damp, cooler, precipitation all seasons.

Mercury-Uranus

Easy, Hard, and Neutral Aspects: descending cold dry air, northwesterly wind with gusts, high pressure, overcast, sudden changes, storms all seasons; hurricanes and tornadoes; hail in warm months.

Hard and Neutral Aspects: major storms when either planet aspected by Venus, Saturn, Neptune, or Mars; gales and strong fronts all seasons.

Mercury-Neptune

Easy Aspects: variable winds, sudden changes, warmer, moderate humidity, fair all seasons.

Hard and Neutral Aspects: high humidity, warm, cloudy, precipitation all seasons; potential for fog in cool months; thaws in cold months; tornadoes and hurricanes.

Mercury-Pluto

Easy Aspects: strong winds all seasons.

Hard and Neutral Aspects: sudden changes, potential for severe wind damage from destructive storms and fronts; hail all seasons.

Venus-Mars

Easy Aspects: fair, warmer all seasons.

Hard and Neutral Aspects: precipitation and storms all seasons.

Venus-Jupiter

Easy and Neutral Aspects: warmer and fair to partly cloudy all seasons.

Hard Aspects: warmer and cloudy all seasons; precipitation.

Venus-Saturn

Easy Aspects: cooler, partly cloudy to fair all seasons.

Hard and Neutral Aspects: overcast, humid/damp, cooler, easterly winds, heavy precipitation, counterclockwise air flow all seasons; rain turning to sleet or snow in winter; fog in autumn and spring.

Venus-Uranus

Easy Aspects: clear and cooler all seasons; cool nights with possible frost; crisp in autumn and spring.

Hard and Neutral Aspects: penetrating cold, overcast, northwesterly winds, and precipitation (drizzle to heavy) all seasons.

Venus-Neptune

Easy Aspects: warmer and humid all seasons; cloudy and showers in summer.

Hard and Neutral Aspects: warmer, heavy precipitation (usually isolated cloudbursts), and flood potential all seasons; thaws in cold months.

Venus-Pluto

Easy Aspects: warmer and breezy all seasons.

Hard and Neutral Aspects: intense precipitation and windy storms all seasons.

Mars-Jupiter

Easy Aspects: warmer and dry, drought all seasons.

Hard Aspects: warmer and storms all seasons.

Neutral Aspects: warmer and thunderstorms all seasons.

Mars-Saturn

Easy Aspects: cloudy and windy all seasons.

Hard and Neutral Aspects: destructive windy storms, followed by cooler temperatures all seasons.

Mars-Uranus

Easy Aspects: gusty breezes, overcast, and cooler all seasons.

Hard and Neutral Aspects: descending cold dry air, and wind and storms all seasons; tornadoes and hurricanes; hail in warm months; sleet in winter.

Mars-Neptune

Easy Aspects: warm and humid all seasons; fog in cooler months.

Hard and Neutral Aspects: humid, rising moist air, and isolated precipitation all seasons; tornadoes and hurricanes; rain and thaw in winter.

Mars-Pluto

Easy Aspects: hot or cold and windy.

Hard and Neutral Aspects: warmer, storms all seasons.

Jupiter-Saturn

Easy Aspects: cooler and possible precipitation all seasons.

Hard and Neutral Aspects: cloudy, storms, low pressure, and below normal temperatures all seasons.

Neutral Aspects: more intense precipitation, low pressure, cooler, and storms all seasons.

Jupiter-Uranus

Easy Aspects: cool, breezy, dry, and fair all seasons.

Hard and Neutral Aspects: cooler, storms, higher precipitation, and windy all seasons.

Jupiter-Neptune

Easy Aspects: above normal temperatures, fair, and sunny all seasons; thaw in winter.

Hard and Neutral Aspects: isolated heavy precipitation with flash flood potential, cloudy, and humid all seasons; mist or fog in cool months.

Jupiter-Pluto

Easy Aspects: unseasonably warm or cold and dry all seasons.

Hard and Neutral Aspects: warm or cold; storms with high winds all seasons.

Saturn-Uranus

Easy Aspects: unseasonably cool all seasons.

Hard and Neutral Aspects: cool and storms all seasons.

Saturn-Neptune

Easy Aspects: precipitation, cooler all seasons.

Hard and Neutral Aspects: above normal precipitation with possible floods and flash floods, humid all seasons; cloudy, night fog in spring; cloudy, heavy precipitation, humid in summer; overcast, chronic heavy rain in autumn; excessive snow, unseasonably cold in winter.

Saturn-Pluto

Easy Aspects: cloudy and windy all seasons.

Hard and Neutral Aspects: storms with high winds all seasons.

Uranus-Neptune

Easy Aspects: unseasonably cool and potential for unusual weather all seasons.

Hard and Neutral Aspects: unseasonably cool, overcast, humid, gusty wind, and storms all seasons.

Uranus-Pluto

Easy Aspects: windy, overcast, and cool all seasons.

Hard and Neutral Aspects: storms with high winds all seasons.

Neptune-Pluto

Easy Aspects: warm, humid, and breezy all seasons.

Hard and Neutral Aspects: warm, humid, wind, and isolated precipitation all seasons.

CHAPTER 3

National and Local Forecasting

You can create a forecast for any location on any date by using basic astrometeorological principles and the ingress and lunar phase charts. Any date can be selected, from tomorrow to ten, fifty, or a hundred or more years in the future. All that's necessary is an ephemeris for the appropriate time frame and charts calculated for the most recent ingress and lunar phase prior to the forecast date. An ephemeris lists the date and time of each solar cardinal ingress, eclipse, and lunar phase, as well as the longitudinal and declination positions of each planet for each day.

You can also generate this information in astrological software programs, many of which can also create an ephemeris and a list of aspects to the lunar phase or ingress chart for a designated time period, as well as a list of transiting planetary aspects.

NATIONAL FORECAST

The ingress chart is used to forecast the general weather conditions for a country or region (as well as for a specific location). The season (spring, winter, autumn, or summer) is determined according to the date for which the ingress chart is calculated. The ingresses, which mark the beginning of each season, occur on or about June 21 (summer),

September 21 (autumn), December 21 (winter), and March 21 (spring). Each is valid for approximately three months, or until the next ingress occurs.

When preparing a national forecast, the goal is to develop a general overview of conditions for the succeeding three months. For example, one possibility for Jupiter square Uranus is an intense cold front. This does not mean the entire country will experience day after day of intense cold fronts, nor does it mean that the area affected by the square will be visited by an intense cold front each day during the forecast period. Rather, temperatures will tend to be below normal for the area where the square is in effect when averaged for the three-month period.

The same can be said, for example, of Venus, a planet of moisture. A location where it is square the meridian might experience an abnormally high level of precipitation, but it would not necessarily rain every day for three months. However, you could correctly forecast above-normal precipitation for the season.

Because the outer planets—Jupiter, Saturn, Uranus, Neptune, and Pluto—move more slowly than the inner planets, they affect long-term, seasonal trends in weather when they aspect the meridian or horizon at a given location. They provide many of the underlying weather trends that are then activated by the inner planets as they transit the zodiac and aspect the ingress and lunar phase charts. An exception to this rule is the retrograde periods of Mercury, Venus, and Mars, as they can have a significant effect on a particular region both by the stations they make and by revisiting certain degrees in certain signs that aspect the meridian or horizon at the lunar phase or ingress.

The easiest way to begin to prepare a national seasonal forecast is to create an ingress map using an astrological mapping program like Solar Maps, included in later versions of Solar Fire. (Version 9 of Solar Fire now simply calls Solar Maps "Astro-Mapping.") The figure shown here (National and Local Forecasting 1) is a computer-generated map for the winter 2011 ingress chart. To generate this map, first calculate the ingress (or lunar phase) chart and then access it in Solar Maps. Use these settings in the Lines menu: Zodiacal (rather than In Mundo) for Planet/Angles lines; do not use Local Space Lines, Celestial Lines, or Paran Lines.

Although the map might appear overwhelming at first, it's really quite straightforward. The aspects to the horizon (As and Ds) and the meridian (Mc and Ic) are shown along the top and bottom and left and right sides of the map. Each aspect has a line drawn to show the locations it aspects.

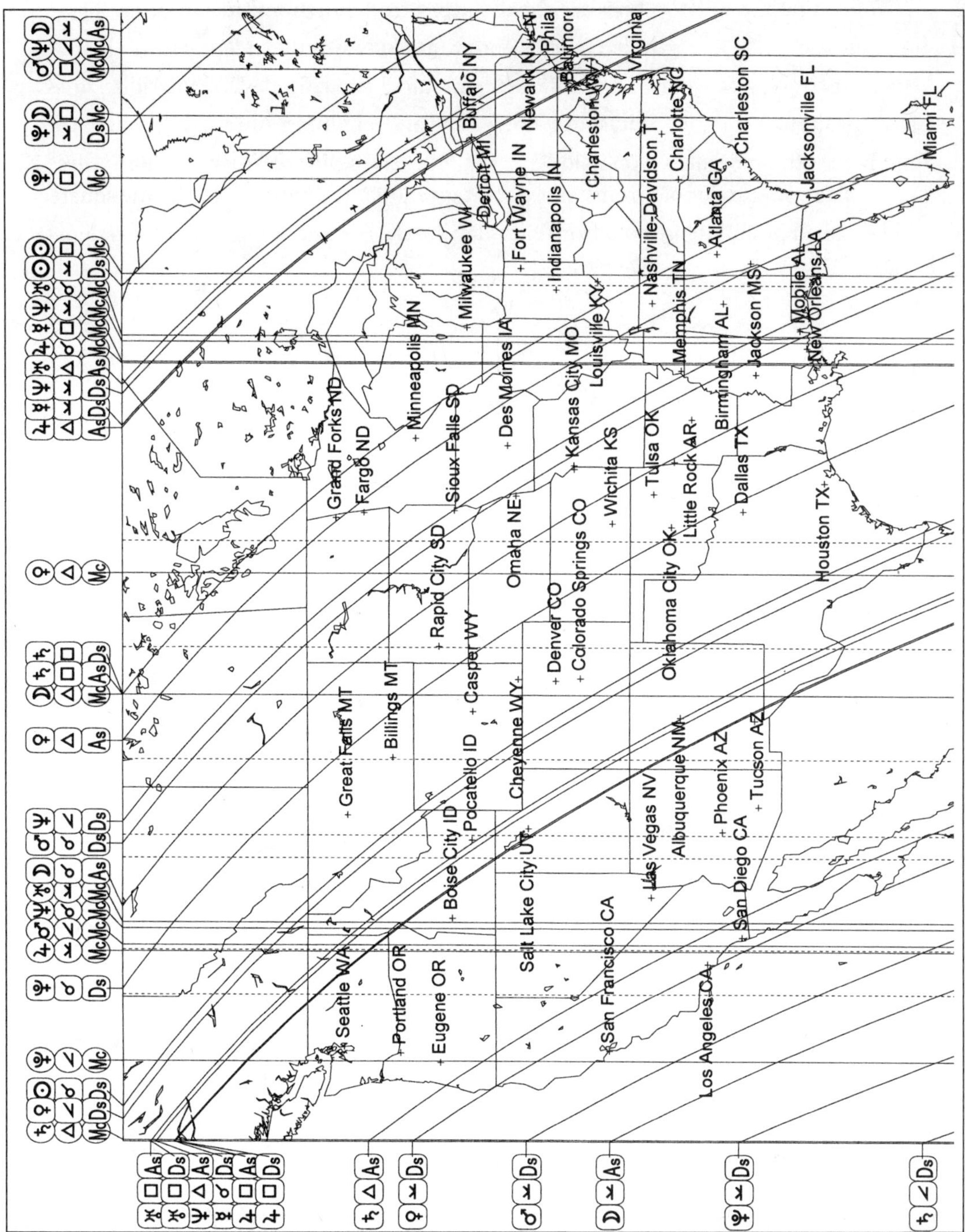

National and Local Forecasting 1: Solar Map for Winter Ingress, December 21, 2010

Notice that the horizon lines are slightly arced from northwest to southeast. These lines often indicate the path that storms and fronts follow as they move across an area, region, or country, aspecting many different longitudes and latitudes. The meridian lines are vertical, running north and south along a specific longitude. Always note where the lines intersect, especially if you're looking for potential severe weather locations; these are "target" locations.

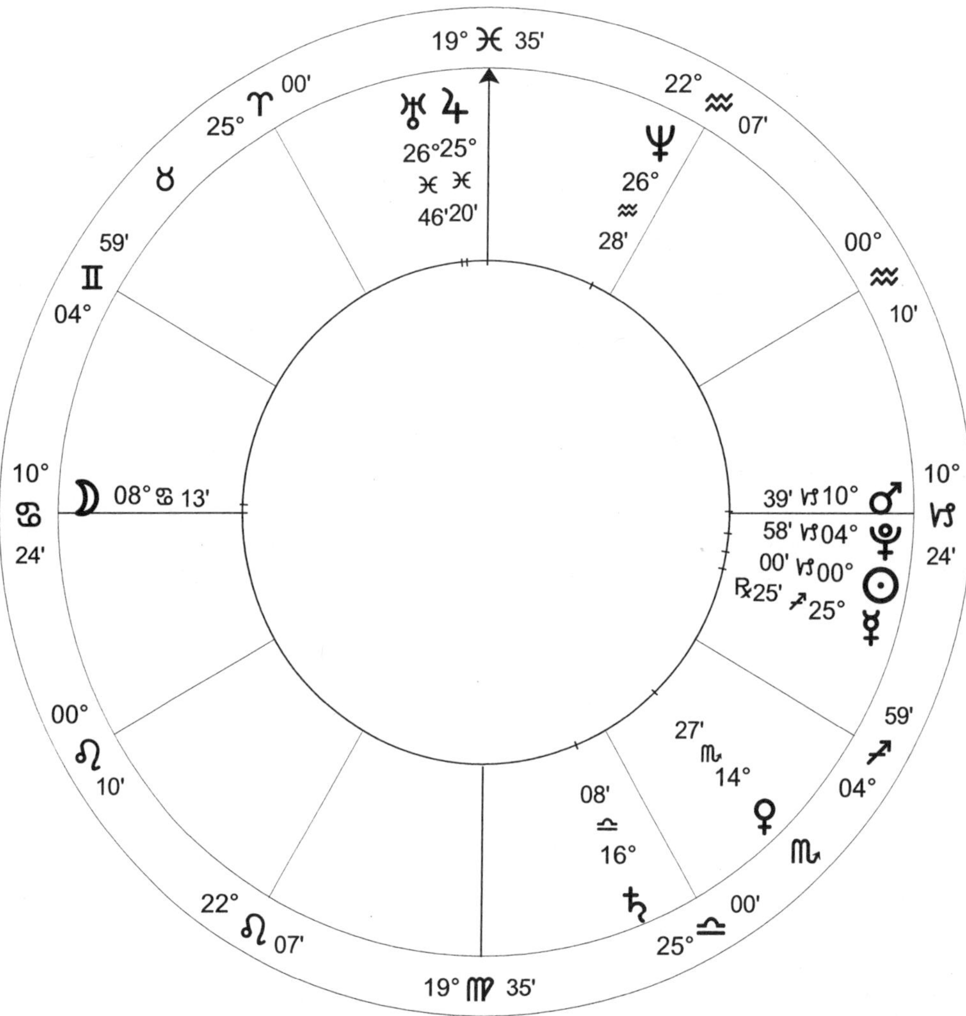

National and Local Forecasting 2: Winter Ingress / Natal Chart / December 21, 2010, Tue
5:40 pm CST +6:00 / 40°N00' 095°W00' / Geocentric / Tropical / Placidus / Mean Node

It doesn't matter what longitude and latitude you use to calculate the chart, because the horizon and meridian lines shown on the map will be visible for all locations. I use a longitude and latitude in the middle of the country (95W 40N) to first calculate the ingress chart; then the mapping program uses this chart to create the map. Using a longitude and latitude for the middle of the country also gives a quick reference in that if you turn the chart clockwise, the planets move to the east, and if you turn it counter-clockwise, they move to the west.

First, let's look at the ingress aspects as shown in the winter ingress chart for 2010 (National and Local Forecasting 2) and what they indicate:

- Sun semisquare Venus: fog, rain, cooler temperatures

- Sun square Uranus: cold, windy, sleet/snow, cold wave

- Sun sextile Neptune: warmer, little wind

- Sun parallel Mars: warmer, windy, storms

- Retrograde Mercury square Jupiter: cloudy, northerly air flow, windy, precipitation

- Retrograde Mercury square Uranus: arctic air mass, high pressure, gusty northwest winds, cold, cloudy, tornadoes, hail

- Retrograde Mercury sextile Neptune: cooler, windy, variable cloudiness, storms, tornadoes

- Venus sextile Mars: fair, warmer, humid/damp; or precipitation

- Venus semisextile Saturn: cooler, fair to partly cloudy

- Venus sesquisquare Uranus: cold, precipitation, cloudy, northwesterly winds, arctic air mass

- Venus parallel Neptune: abundant precipitation, cloudy, warmer

- Mars semisquare Neptune: thaw, warmer, fog, low pressure, tornadoes

- Jupiter conjunct Uranus: high pressure, cold, cold front, precipitation, windy, arctic air mass

- Jupiter semisextile Neptune: warmer, fair, thaw

- Solar eclipse: January 4, 2011, at 13 Capricorn

As is obvious from this list, the weather could be almost anything and everything. So the next task is to determine which weather pattern will dominate in which location. For that, we use the map (National and Local Forecasting 1) and its aspects.

Before analyzing the map, though, there are two important considerations to remember. First, consider the local climatology. Second, the weather is never the same in any location day after day throughout an entire season. For example, if the ingress map shows heavy precipitation in Denver, that doesn't mean it will snow every day of the season. The ingress chart shows trends and the potential for specific weather events.

Now look at the Uranus conjunct Mc line, which is just to the right of the Mercury square Mc line. Because both planets aspect approximately the same longitude, temperatures will be below average in these states (very cold in northern locations), and there is high potential for severe thunderstorms in southern locations.

The Saturn square Asc line represents the path storms will follow from Montana southeast to Georgia/Florida, and it crosses the Mercury and Uranus lines in central Illinois. The intersection of these lines also reinforces the already present indication of below average temperatures. The Mars conjunct Ds, Neptune semisquare Ds, and Venus trine As lines (look above Montana) also intersect the Mercury and Uranus lines. On their own, these planets indicate strong thunderstorms, with Venus providing moisture; in combination with the Mercury and Uranus lines, severe storms with tornado potential are likely.

In January 2011, there were tornadoes in Mississippi. January also brought a cold wave that dropped the temperature to -46 F at International Falls, Minnesota, and later in the month, frigid temperatures into the Deep South (ingress Pluto was square the meridian and Saturn was aligned in hard aspect with the horizon). The January solar eclipse was also active in this area because it was aligned with the latitude line that ran from Wisconsin southeast to Alabama.

Now look at the West, where Neptune is conjunct Mc (look above Idaho/western Montana) along a longitude that runs through the mountains. Venus semisquare Ds intersects the Neptune line, as do retrograde Mercury and Jupiter. This region received record snowfall during the winter of 2011. Notice the other latitude lines to the east of there: Mars, Neptune, Venus, and Saturn. Storms traveled from the West into the Midwest, including a blizzard at the end of January/early February, and a major storm in Oklahoma and Arkansas, followed by extreme cold, dropping the temperature to -31 F in Oklahoma (ingress Pluto was aligned in hard aspect with the horizon).

It's also important to look at the signs on the horizon and meridian, although planetary aspects take priority. Finding the signs on the meridian requires some mental interpolation because the signs are not shown on the map. This is where the ingress chart calculated for the middle of the country comes in handy. When in doubt, calculate a chart for several regions in order to determine where the longitudinal sign changes.

Each degree of longitude (or latitude) equals about 1 degree of a sign. So if you begin at the middle of the country where 19 Pisces is on the Midheaven, you know that the Midheaven sign will change to Aquarius 19 degrees to the west, or at about 105W (eastern Montana south to New Mexico). To the east, the Midheaven sign will change to Aries about 11 degrees to the east, or at about 85W (Michigan south to Alabama). You can identify the latitude signs based on the planetary latitude lines because the chart tells you what sign the planet is in; use the same method of decreasing or increasing the sign degrees based on the degrees of latitude. It can be helpful to write the signs and the declinations along the top of the map and along the latitude lines.

You can also find the Ascendant by using a tables of houses. To do this, first find the page that has the appropriate Midheaven degree and sign listed at the top. Then move down the center column to the correct longitude and look to the left or right to find the sign degree under the Ascendant column.

However, I often find it easier to use an old-fashioned technique: a hand-drawn map. If you want to do this, use a country map with state or province outlines and write the Midheaven and Ascendant signs on it along with the ingress planets that aspect them.

Sign interpretation can be a bit of challenge because you're dealing with two signs for the horizon—one on the Ascendant and one on the Descendant—and two signs for the meridian—one on the Midheaven and one on the IC. Somewhere in the "middle," the sign influence will flip from one to the other. For example, in the winter of 2010 ingress chart (National and Local Forecasting 2), the Sun at 0 Capricorn is conjunct the latitude line at about 105W 40N, which you can determine by rotating the chart counterclockwise until you reach the approximate Midheaven of 9 Pisces and then mentally adjusting the latitude degrees. (Keep in mind that this method is not perfect; it will, however, give you a general idea, which you can always confirm by calculating another chart.)

So, northwest of the 105W 40N, where Cancer is emphasized, temperatures should generally be warmer than those to the southeast, where Capricorn is emphasized. This was true because the southern states had far colder weather than the northern ones during that winter. You can do the same with the latitude lines, remembering that the

Midheaven is south and the IC is north (in the Northern Hemisphere). In much of this ingress chart, the Pisces (wet) influence was to the south and the Virgo (dry) influence to the north, which also characterized the winter weather.

Again, though, the sign influence pales in comparison to the planetary influence. It's far more influential when a wet planet is in a wet sign, a dry planet in a dry sign, a hot planet in a hot sign, etc.; at this ingress, Venus, a planet of precipitation, was in wet Scorpio.

It's also a good idea to scan the ephemeris for the succeeding three months to determine when the inner planets will transit the signs and degrees of the ingress chart planets, because these transits often trigger weather. It might not be as accurate or as detailed a forecast as one that incorporates the use of the lunar phase chart, but it nevertheless works fairly well and also can alert you to look at the lunar phase chart for confirmation. Do the same with the ingress chart horizon and meridian degrees and signs. For example, you can spot a heat wave by tracking the aspects of transiting Mars, particularly the conjunction or square, to the ingress meridian degrees and signs. When a Mercury or Venus retrograde or direct station aspects an ingress (or lunar phase) meridian, it can indicate a stationary front (a front that is "parked" or moves very slowly and often brings high levels of precipitation).

January 9–12, 2011, provides a good example to illustrate the effects of transiting planets to the ingress chart. Transiting Mars moved from 24 to 27 Capricorn and Venus was at 2–6 Sagittarius on those dates, as transiting Saturn moved toward a retrograde station at 17 Libra—which you can see by quickly scanning the ephemeris shown here (National and Local Forecasting 3). During these few days, a storm moved from the South to the Northeast along the East Coast of the United States.

Transiting Saturn aspected the ingress horizon in Georgia and the ingress meridian in the Northeast, transiting Mars aspected the ingress horizons from South Carolina to New York, and transiting Venus aspected many of the same locations through semisquare aspect. Transiting Mars was also sextile a transiting Jupiter-Uranus conjunction and semisextile transiting Neptune. Just from these aspects you would know to check the lunar phase charts for more information. Note also how the transiting planets indicate the direction in which the storm traveled. The ingress Ascendant in the South was 24 Capricorn, and in New York it was 27 Capricorn. The ingress Midheaven in the Northeast was 17 Aries, and Venus aspected the ingress horizons and meridians by sesquisquare (17–21 Capricorn) and the ingress latitude in many Southern locations by trine. Note that the storm traveled the path of the hard aspects rather than the easy one (trine).

04 08:52 13♑39 ☉ Solar Eclipse (mag 0.858)

January 2011

Day	S. T. (h m s)	☉	☽	☿	♀	♂	♃	♄	♅	♆	♇	☊ True
01 Sa	06 41 13	10♑12 10	29♏13 53	19♐50	23♏34	18♐22	26♓33	16♎40	26♓58	26♒44	05♑20	02♑42
02 Su	06 45 09	11 13 20	12 38 24	20 09	24 32	19 09	26 41	16 42	26 59	26 46	05 22	02 43
03 Mo	06 49 06	12 14 31	25 50 11	20 35	25 30	19 55	26 49	16 45	27 00	26 48	05 24	02 43
04 Tu	06 53 02	13 15 42	08♒48 38	21 08	26 29	20 42	26 58	16 47	27 02	26 49	05 26	02R 43
05 We	06 56 59	14 16 53	21 33 25	21 46	27 28	21 28	27 06	16 49	27 03	26 51	05 28	02 43
06 Th	07 00 56	15 18 03	04♒04 45	22 30	28 28	22 15	27 15	16 52	27 05	26 53	05 30	02 42
07 Fr	07 04 52	16 19 14	16 23 28	23 19	29 28	23 02	27 24	16 54	27 06	26 55	05 33	02 41
08 Sa	07 08 49	17 20 24	28 31 10	24 12	00♐29	23 48	27 33	16 56	27 08	26 57	05 35	02 39
09 Su	07 12 45	18 21 34	10♓30 11	25 09	01 30	24 35	27 42	16 58	27 10	26 58	05 37	02 38
10 Mo	07 16 42	19 22 43	22 23 34	26 09	02 32	25 22	27 51	16 59	27 11	27 00	05 39	02 36
11 Tu	07 20 38	20 23 52	04♈15 06	27 12	03 33	26 08	28 01	17 01	27 13	27 02	05 41	02 36
12 We	07 24 35	21 25 00	16 09 06	28 18	04 36	26 55	28 10	17 03	27 15	27 04	05 43	02D 36
13 Th	07 28 31	22 26 08	28 10 16	29 27	05 38	27 42	28 20	17 04	27 17	27 06	05 45	02 36
14 Fr	07 32 28	23 27 15	10♉23 22	00♑37	06 41	28 29	28 30	17 06	27 19	27 08	05 47	02 38
15 Sa	07 36 25	24 28 22	22 53 01	01 50	07 44	29 16	28 40	17 07	27 20	27 10	05 49	02 40
16 Su	07 40 21	25 29 28	05♊43 12	03 04	08 48	00♒03	28 50	17 08	27 22	27 12	05 52	02 42
17 Mo	07 44 18	26 30 33	18 56 54	04 20	09 52	00 49	29 00	17 09	27 24	27 14	05 54	02 43
18 Tu	07 48 14	27 31 38	02♋35 33	05 37	10 56	01 36	29 10	17 10	27 27	27 16	05 56	02R 43
19 We	07 52 11	28 32 42	16 38 27	06 56	12 00	02 23	29 20	17 11	27 29	27 18	05 58	02 42
20 Th	07 56 07	29 33 46	01♌02 28	08 16	13 05	03 10	29 31	17 11	27 31	27 20	06 00	02 39
21 Fr	08 00 04	00♒34 48	15 42 05	09 37	14 10	03 57	29 42	17 12	27 33	27 22	06 02	02 35
22 Sa	08 04 00	01 35 50	00♍30 05	10 59	15 15	04 44	29 52	17 13	27 35	27 24	06 04	02 31
23 Su	08 07 57	02 36 52	15 18 31	12 22	16 20	05 31	00♈03	17 13	27 37	27 26	06 06	02 26
24 Mo	08 11 54	03 37 53	29 59 58	13 47	17 26	06 19	00 14	17 13	27 40	27 29	06 08	02 21
25 Tu	08 15 50	04 38 54	14♎28 30	15 12	18 32	07 06	00 25	17 14	27 42	27 31	06 10	02 17
26 We	08 19 47	05 39 54	28 40 15	16 38	19 38	07 53	00 36	17 14	27 44	27 33	06 12	02 15
27 Th	08 23 43	06 40 54	12♏33 29	18 04	20 45	08 40	00 47	17R 14	27 47	27 35	06 13	02 14
28 Fr	08 27 40	07 41 53	26 08 14	19 32	21 51	09 27	00 59	17 13	27 49	27 37	06 15	02D 14
29 Sa	08 31 36	08 42 52	09♐25 45	21 00	22 58	10 14	01 10	17 13	27 52	27 39	06 17	02 15
30 Su	08 35 33	09 43 50	22 27 52	22 30	24 05	11 02	01 22	17 13	27 54	27 41	06 19	02 16
31 Mo	08 39 29	10 44 47	05♑16 30	23 59	25 12	11 49	01 33	17 12	27 57	27 44	06 21	02R 16

Data for 01-01-2011
Julian Day 2455562.50
Ayanamsa 24 00 55
SVP 05♓06 40
☽ ☊ Mean 02♑18 R

● ◐ PHASES ○ ◑

04	09:03 ☉	13♑39
12	11:31 ◐	21♈54
19	21:22 ○	29♋27
26	12:58 ◑	06♏13

LAST ASPECT ☽ INGRESS

Day h m	Day h m	
31 19:58	01 01:22	♐
03 02:09	03 07:39	♑
05 12:16	05 16:08	♒
07 20:51	08 02:57	♓
10 11:12	10 15:24	♈
13 02:47	13 03:37	♉
15 12:47	15 13:23	♊
17 17:58	17 19:30	♋
19 21:27	19 22:17	♌
21 18:59	21 23:11	♍
23 20:09	24 00:00	♎
25 22:05	26 02:16	♏
28 03:01	28 06:55	♐
30 10:11	30 14:04	♑

DECLINATION

Day	☉	☽	☿	♀	♂	♃	♄	♅	♆	♇
01 Sa	23S03	22S50	20S13	15S17	23S10	02S32	04S19	01S53	13S03	18S50
02 Su	22 58	24 07	20 21	15 30	23 03	02 29	04 20	01 53	13 03	18 50
03 Mo	22 52	24 00	20 30	15 44	22 57	02 25	04 21	01 52	13 02	18 50
04 Tu	22 47	22 35	20 39	15 57	22 50	02 22	04 21	01 51	13 02	18 50
05 We	22 40	20 02	20 50	16 10	22 43	02 18	04 22	01 51	13 01	18 50
06 Th	22 34	16 34	21 01	16 24	22 36	02 14	04 23	01 50	13 00	18 50
07 Fr	22 26	12 27	21 12	16 37	22 28	02 11	04 23	01 49	13 00	18 50
08 Sa	22 19	07 54	21 24	16 50	22 21	02 07	04 24	01 49	12 59	18 50
09 Su	22 11	03 07	21 35	17 03	22 13	02 03	04 24	01 48	12 58	18 49
10 Mo	22 02	01N44	21 46	17 16	22 04	01 59	04 25	01 47	12 58	18 49
11 Tu	21 53	06 30	21 57	17 28	21 56	01 55	04 25	01 47	12 57	18 49
12 We	21 44	11 03	22 07	17 41	21 47	01 51	04 25	01 46	12 56	18 49
13 Th	21 34	15 14	22 17	17 53	21 38	01 47	04 26	01 45	12 56	18 49
14 Fr	21 24	18 52	22 26	18 05	21 29	01 43	04 26	01 44	12 55	18 49
15 Sa	21 13	21 44	22 34	18 17	21 19	01 39	04 26	01 43	12 54	18 49
16 Su	21 02	23 36	22 42	18 28	21 10	01 35	04 26	01 43	12 54	18 49
17 Mo	20 51	24 13	22 48	18 39	21 00	01 31	04 26	01 42	12 53	18 49
18 Tu	20 39	23 25	22 54	18 50	20 49	01 26	04 27	01 41	12 52	18 49
19 We	20 27	21 08	22 59	19 01	20 39	01 22	04 27	01 40	12 52	18 49
20 Th	20 14	17 29	23 03	19 11	20 28	01 18	04 27	01 39	12 51	18 49
21 Fr	20 02	12 43	23 05	19 21	20 18	01 13	04 27	01 38	12 50	18 49
22 Sa	19 48	07 09	23 07	19 30	20 06	01 09	04 27	01 37	12 50	18 49
23 Su	19 34	01 12	23 07	19 40	19 55	01 04	04 26	01 37	12 49	18 49
24 Mo	19 20	04S46	23 07	19 48	19 43	01 00	04 26	01 36	12 48	18 49
25 Tu	19 06	10 24	23 05	19 57	19 32	00 55	04 26	01 35	12 47	18 49
26 We	18 51	15 21	23 02	20 05	19 20	00 51	04 26	01 34	12 47	18 49
27 Th	18 36	19 23	22 58	20 12	19 08	00 46	04 26	01 33	12 46	18 49
28 Fr	18 21	22 16	22 52	20 20	18 55	00 41	04 25	01 32	12 45	18 49
29 Sa	18 05	23 52	22 45	20 26	18 42	00 37	04 25	01 31	12 44	18 49
30 Su	17 49	24 06	22 37	20 33	18 30	00 32	04 25	01 30	12 44	18 49
31 Mo	17 32	23 03	22 28	20 38	18 17	00 27	04 24	01 29	12 43	18 48

⚷ Chiron

Day		
01	Dec.	06 S 54
03	27♒43	
06	27 53	
09	28 03	
12	28 13	
15	28 24	
18	28 35	
21	28 46	
24	28 57	
27	29 09	
30	29 21	

ASPECTARIAN

```
01 02:31  ☽ ‖ ☉          16 06:08  ☽ ☍ ♂      19 00:54  ☽ □ ♄      08:57  ☽ ‖ ♃      11:05  ☽ ‖ ♆      03:01  ☽ △ ♅
   03:56  ☽ ‖ ♂             20:47  ☽ △ ♄         04:02  ☽ ☍ ♂      11:10  ☽ ‖ ♅      22:05  ☽ △ ♆      06:31  ☽ ‖ ♅
02 07:22  ☽ ✶ ♄         17 14:42  ☽ △ ♅         05:39  ☽ ☍ ☿      18:08  ♀ ‖ ♂                        08:49  ☽ △ ♆
   14:03  ☽ ☌ ☿             15:00  ☽ □ ♂         14:21  ☽ △ ♀      19:19  ♀ ✶ ♄                       12:18  ♂ ‖ ♆
03 01:45  ☽ ✶ ♀             17:58  ☽ ♃ ♄         16:12  ☽ △ ♇      20:09  ☽ ♃ ♆   26 03:53  ☉ ‖ ♆      22:35  ☽ ✶ ☉
   01:50  ☽ □ ♄             21:57  ☉ ✶ ♃         18:09  ☽ ☍ ♃      23:38  ☽ ‖ ♄         06:11  ♄ SR  29 01:34  ☽ ✶ ♅
   02:09  ☽ □ ♅             21:58  ☽ ‖ ♂         21:27  ☽ △ ♃                             09:58  ♂ ✶ ♃      14:17  ☽ ✶ ♅
   17:42  ☽ ☌ ♆      06:26  ☽ ✶ ♂     20:01  ☽ ‖ ☉         22:39  ☉ ✶ ♃   22 09:01  ☽ △ ♃         12:58  ☽ ✶ ♆
   20:48  ☽ ‖ ♂      08:20  ☽ □ ♄         20:04  ☽ ✶ ♅   20 03:43  ☽ △ ♆         11:02  ☽ ‖ ♄         16:49  ♃ □ ♂  30 03:18  ☽ ☌ ♂
   21:37  ☽ ‖ ☉      09:44  ☽ □ ♅         22:28  ☽ △ ♇         10:19  ☉ ‖ ♆         17:11  ♃ ♈         19:06  ☽ ‖ ☉      09:46  ☽ ✶ ♅
04 08:28  ♀ □ ♆      11:12  ☽ ☌ ♃         23:06  ☽ ‖ ♂         21:18  ☽ △ ♇         18:34  ☽ □ ♅         20:08  ☽ ☍ ♃      10:11  ☽ □ ♀
   12:52  ♃ ☌ ♆      13:23  ☽ ✶ ♄  11 00:13  ☽ □ ♅         23:26  ☽ ✶ ♆         18:44  ☽ △ ♀         22:18  ☽ ‖ ♂      16:53  ☽ □ ♃
   13:34  ♀ ☌ ♇      20:01  ☽ ‖ ☉         02:54  ☽ □ ♆   21 02:26  ☽ ✶ ♄         21:21  ☽ ‖ ♃   24 00:23  ☽ ☍ ♃  31 02:02  ☽ □ ♀
   13:40  ♀ △ ♃      20:04  ☽ ✶ ♅         20:37  ☽ ‖ ♂      18:59  ☽ ☍ ♀   23 00:30  ☽ ‖ ♄         06:26  ☽ △ ☉      08:37  ☽ ‖ ♄
   15:01  ☽ □ ♄      22:28  ☽ △ ♇  12 01:48  ☽ ☍ ♆                             01:49  ☽ △ ♀         10:08  ☽ □ ♀      22:40  ☽ □ ♄
   17:49  ☽ ‖ ♂      23:06  ☽ ‖ ♂         10:26  ☽ ☍ ♄   22 09:01  ☽ △ ♃                             11:00  ☽ △ ♇
   23:50  ☽ ☌ ♂                           10:30  ♂ ✶ ♅                         25 01:20  ☽ □ ♄   27 06:06  ☽ □ ♄
05 08:59  ☽ ☌ ♆                           10:43  ☽ ✶ ♃                             04:37  ☽ ☌ ♀      10:51  ☽ ✶ ♀
   10:30  ☽ ✶ ♅                           12:16  ☽ ♃ ♀                             07:23  ☽ ✶ ♀   28 02:39  ☽ □ ♆
   10:43  ☽ ✶ ♃                    13 02:47  ☽ △ ♀
   12:16  ☽ ☌ ♀                           11:25  ☉ ♑
06 01:01  ☽ ‖ ♇                           15:00  ☽ △ ♅
   20:59  ☽ ‖ ♄                           15:21  ♃ ‖ ♅
07 01:00  ☽ △ ♄                           18:00  ♀ ‖ ♂
   12:31  ♀ ☌ ♇                           23:36  ☽ ♃ ♀
   14:01  ☉ □ ♄                    14 00:36  ☽ ✶ ♃
   14:44  ☽ ✶ ♅                           19:21  ☽ ♃ ☉
   20:51  ☽ ☌ ♆                           20:11  ☽ ♃ ♀
08 04:17  ☽ □ ♀                    15 03:16  ☽ △ ☉
   14:09  ☽ ✶ ♀                           08:06  ☽ □ ♅
   17:37  ☽ ‖ ♄                           08:26  ☽ ♃ ☿
   17:20  ☽ ✶ ☉                           09:35  ☽ ♃ ♀
09 05:22  ☽ ‖ ♄                           11:02  ☽ ✶ ♂
   06:32  ☽ ‖ ♂                           12:47  ☽ △ ♂
   17:20  ☽ ✶ ☉                           22:42  ♂ ♒
10 00:16  ☽ ♃ ♆
   01:14  ☽ ♃ ♃
```

National and Local Forecasting 3: January 2011 Ephemeris Page

Local Forecasting

Local forecasting is tough and can really test your forecasting skills, particularly if you're looking at a specific date. It requires close examination of the ingress and lunar phase charts as well as the transits, and even then it's easy to miss the triggering aspect or to be off on the timing, as this example for Long Pond, Pennsylvania, on August 7, 2011, illustrates (National and Local Forecasting 4); rain began at 3:20 pm.

The summer ingress chart for Long Pond (inner wheel) had several planetary configurations that indicated rain and storm potential during the season. Most obvious is the Uranus-Pluto square, with Pluto aligned with the meridian and Uranus aligned with the horizon and meridian, along with the Sun conjunct the Midheaven, creating a T-square configuration. In itself, Uranus square the meridian is a thunderstorm and weather front aspect, and it's even stronger because Saturn was contraparallel Uranus (storms).

Also in this chart was a Mercury-Saturn square, which was widely in aspect to the Uranus-Pluto square, but neither Mercury nor Saturn aspected the meridian or horizon. Mars was trine/sextile the horizon and semisextile/inconjunct the meridian. The Sun was parallel Venus, and Venus was parallel the Midheaven and Ascendant, which indicated an increased potential for moisture in Long Pond.

At the lunar phase (middle wheel of National and Local Forecasting 4), Uranus was still square Pluto, with both planets aspecting the meridian (semisquare/sesquisquare). Mars was conjunct the ingress Midheaven; it was also semisquare/sesquisquare the lunar phase meridian, and square Uranus, opposition Pluto, and trine Neptune, which was opposition retrograde Mercury (rain). Mars was sextile retrograde Mercury (wind), also aspecting the ingress meridian and horizon.

Also at the lunar phase, the Sun was conjunct Venus (rain), which was square Jupiter (rain) and also sextile Saturn (cloudy skies), which was in turn sesquisquare Neptune (rain). These aspects gain added emphasis because the Sun and Venus were parallel and both were contraparallel Pluto; all three were in declination aspect with the lunar phase meridian.

Nearly every aspect at the lunar phase forecasted rain, and this only increased with the transits of August 7 (outer wheel of National and Local Forecasting 4). Retrograde Mercury was still opposition Neptune and in aspect with the ingress meridian and horizon. Venus had advanced to a nearly exact sextile with lunar phase Saturn and was still parallel the lunar phase Midheaven, activating that point.

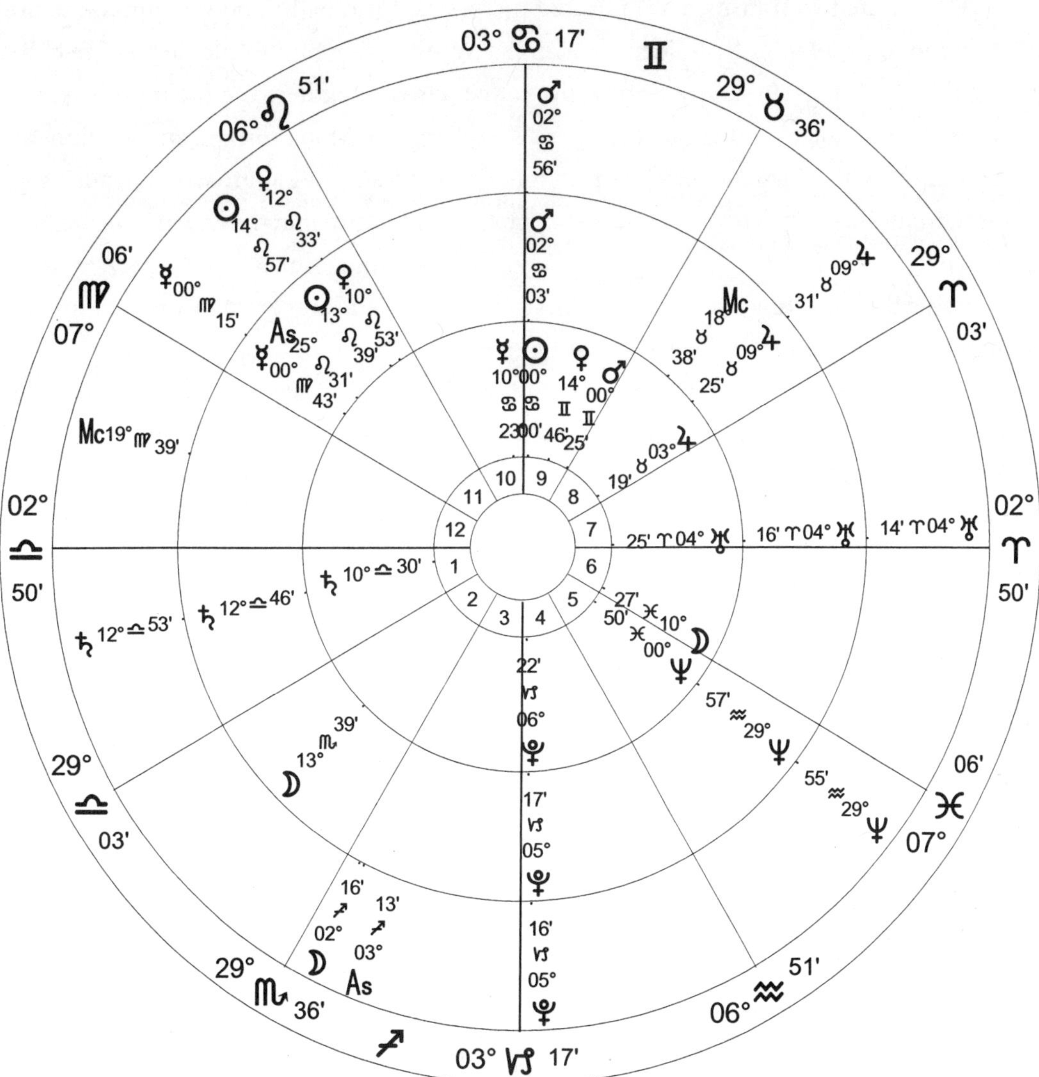

National and Local Forecasting 4

Inner Wheel=*Summer Ingress / Natal Chart / June 21, 2011, Tue / 1:18 pm EDT +4:00*
Long Pond, PA / 41°N03'12" 075°W27'48" / Geocentric / Tropical / Placidus / Mean Node

Middle Wheel=*Lunar Phase / Natal Chart / August 6, 2011, Sat / 7:08 am EDT +4:00*
Long Pond, PA / 41°N03'12" 075°W27'48" / Geocentric / Tropical / Placidus / Mean Node

Outer Wheel=*Rain / Natal Chart / August 7, 2011, Sun / 3:20 pm EDT +4:00*
Long Pond, PA / 41°N03'12" 075°W27'48" / Geocentric / Tropical / Placidus / Mean Node

There were two transits that indicated the timing for rain, but both required much attention to detail in order to identify within a minute when rain would begin. The first was the transiting Moon at 2 Sagittarius 16, which was sextile/trine the ingress horizon and inconjunct/semisextile the ingress meridian. The Moon formed an inconjunct/semisextile with Mars and was just past lunar phase Mars and approaching transiting Mars. All of these aspects were close but not exact, and notice that the Moon had already passed its aspect to retrograde Mercury opposition Neptune, a major rain aspect.

It was the transiting Ascendant at 3 Sagittarius 13 in a nearly exact inconjunct to the ingress Midheaven that indicated the timing; this was the triggering aspect.

Much easier than forecasting the onset of weather within a few minutes is to identify a time frame. It was clear that there would be rain on August 7, 2011, and you could have done well by selecting a window of about an hour from 3:00 to 4:00 pm as the Moon formed an exact aspect with the lunar phase Mars-Midheaven conjunction and the ingress Ascendant. That would have been more than adequate for most events, and you could have refined the timing on the go by checking local radar.

CHAPTER 4

Wind

From winter blasts of frigid northern air to the gentlest of summer breezes, wind is often associated with changing weather patterns. Wind is created by tight pressure gradients, which are shown on weather maps. They're the curving lines drawn around the center of a high- or low-pressure system or in the vicinity of a weather front. When there are many lines close together, the wind velocity is higher; when there are fewer lines, the wind velocity is lower. High pressure pushes air toward low pressure, and wind velocity increases either as the two systems move closer together or as the contrast in pressure between the high- and low-pressure systems increases.

Barometric pressure typically ranges from 980 mb (very low) to 1049 mb (very high), but there is no cutoff number to determine high or low pressure. Wind velocity is also determined by friction as the air moves around buildings, trees, and geologic formations such as hills and mountains. There is much less friction over water than over land, because there is little to slow the movement of air. The Coriolis effect (related to the spinning of Earth) also contributes to wind velocity.

Air moves clockwise around high-pressure systems and counterclockwise around low-pressure systems. High pressure generally results in favorable weather, while low pressure is associated with stormy conditions.

In a high-pressure system, descending cold air is warmed by the surrounding higher-pressure air. Water in the air evaporates and becomes vapor, but the warming keeps the vapor from condensing into clouds and precipitation. The opposite is true with a low-pressure system. The rising air cools and condenses water vapor into clouds and precipitation. The greater the pressure differences between the high- and low-pressure systems, the greater the temperature differences between two locations. You've probably experienced this when a cold front moves through and the temperature drops significantly.

Earth is a complex weather machine with major pressure patterns around the globe, such as the Icelandic Low, Aleutian Low, Asian Highlands, Bermuda High, and Pacific High. These pressure patterns, global wind patterns, and temporary, constantly changing high- and low-pressure systems are central to daily weather. Wind patterns include the jet stream, which has a significant effect in moving weather around the globe.

The jet stream travels from west to east in bands of high-speed, high-altitude (250 mb or 35,000 feet) wind that range from 80 to more than 170 mph. Swooping north and south and curving around massive high- and low-pressure systems, the jet stream defines the boundaries between warmer and colder air on the earth's surface. In winter months, the jet stream drops farther south, bringing cold Arctic air; in the summer, it recedes to more northern latitudes. However, when the jet stream drops far south in warmer months, temperatures are colder than normal; when it travels in a more northerly pattern during the winter months, southern (and sometimes northern) temperatures are warmer than normal. The strongest winds within the jet stream are found in jet streaks, areas where the upper air converges or diverges, resulting in cyclonic intensification (divergence and lowering surface pressure) or weakening (convergence and rising surface pressure).

Jupiter and Uranus are the planets of high pressure; Saturn and Neptune signify low pressure. Uranus dominates when cold air descends from the upper atmosphere.

Mercury, the most prominent planet of wind, manifests according to the character of the planets it aspects. However, other planets—Mars, Uranus, and Pluto—and planetary configurations also indicate wind. When any of these planets (Mercury, Mars, Uranus, Pluto) are on the angles or square the meridian, the forecast includes wind, the intensity and nature of which depends on the aspects they make and which planets are involved.

Wind direction changes depending on which planets are active. The general wind directions associated with the planets are as follows: southerly, Venus and Neptune; northerly, Jupiter; northwesterly, Uranus; easterly, Saturn; westerly, Mars and Pluto; and variable, Neptune. Uranus indicates downslope winds, while Neptune is associated with an upslope air flow.

A counterclockwise air flow, such as that found in tornadoes, extratropical cyclones, and hurricanes, is reflected in aspects with Saturn, Neptune, or Uranus. Venus and Neptune indicate breezes and Uranus signals erratic gusts. Mars and Pluto are associated with higher velocities, while Neptune can indicate little or no air movement. Retrograde Mercury, especially when stationary direct or retrograde and strongly placed in an ingress or lunar phase chart, often indicates high wind.

In this chapter, we'll look at the planetary configurations as they relate solely to wind. Wind as a storm component is included in succeeding chapters.

DECEMBER 14, 2002, RENO, NEVADA

Reno, Nevada, experienced winds ranging from 24 to 60 mph on December 14, 2002, including an 82 mph gust, the strongest ever recorded in the area. The winds, which were at their fastest between 3:00 and 4:00 pm, blew shingles off roofs and toppled trucks, power lines, billboards, carports, trees, and fences.

This wind event is a good example of the power of retrograde Mercury. The planet of wind was retrograde at 9 Libra at the autumn ingress (inner wheel of Wind 1), where it was sextile Jupiter and the IC. Also important was Mercury's sesquisquare to Uranus, a high-wind and gust aspect, particularly with retrograde Mercury. In addition, Saturn was trine Uranus and semisquare Jupiter and the IC, bringing in the potential for the lower barometric pressure associated with tight pressure gradients.

On its own, the lunar phase chart of December 11, 2002 (middle wheel of Wind 1), shows maximum potential for a weather event. Mars (wind) was conjunct Venus and the Midheaven at 4–7 Scorpio, and all were square Neptune. Venus and Neptune are both associated with a southerly air flow, and winds blew from the south all day. In addition, Venus, Mars, and Neptune indicate warmer temperatures; Reno had a high that day of 63 F, almost 20 F above the average for December 14. The same date saw the jet stream far north over the northwestern United States, reflecting the warmer temperatures in Reno.

Another clear sign that there would be a wind-related weather event in Reno was the lunar phase Ascendant at 11 Capricorn. Ingress retrograde Mercury was square and Uranus sesquisquare this point. And, of course, the lunar phase Ascendant was semisextile/inconjunct the ingress Midheaven/IC-Jupiter, linking the two charts. But even this was not enough to suggest the high-wind emergency that occurred, or at least not its timing. So we look to the transits of December 14, the date of the event (outer wheel of Wind 1).

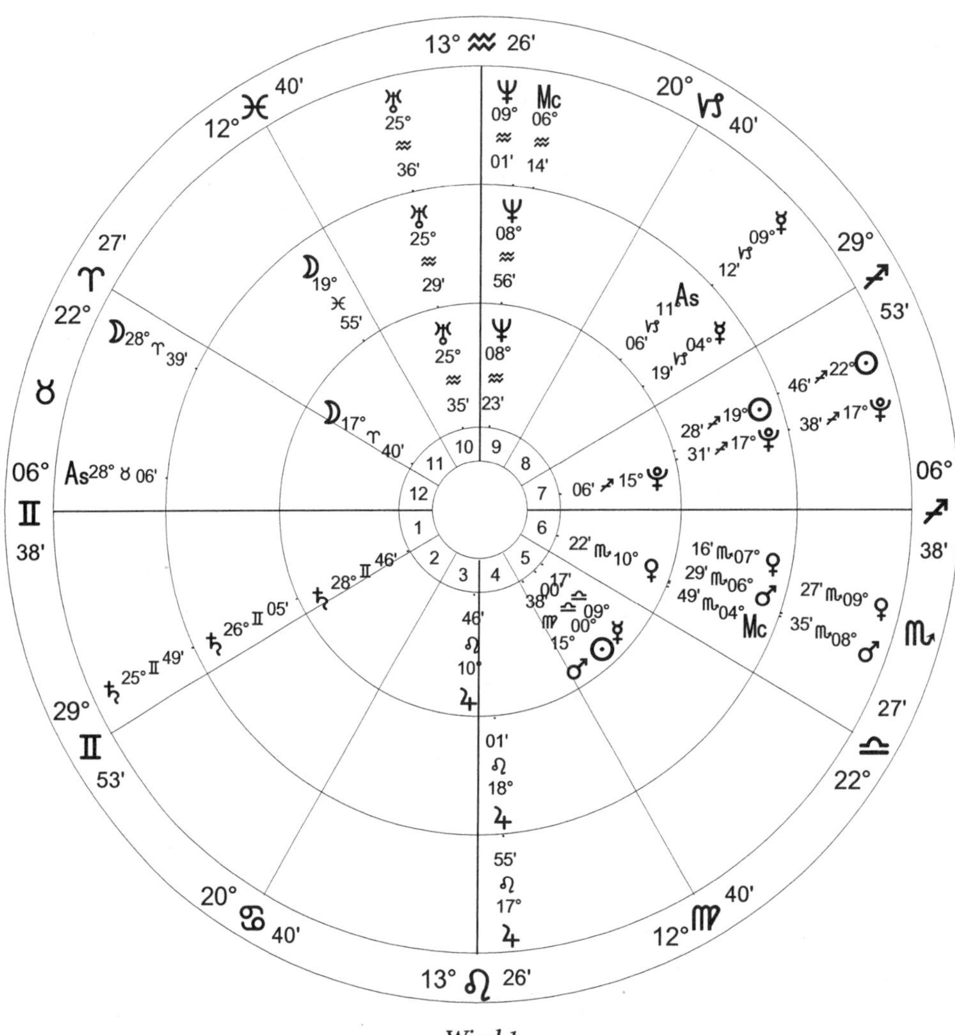

Wind 1

INNER WHEEL=*Autumn Ingress / Natal Chart / September 22, 2002, Sun / 9:55:24 pm PDT +7:00 Reno, NV / 39°N31'47" 119°W48'46" / Geocentric / Tropical / Placidus / Mean Node*

MIDDLE WHEEL=*Lunar Phase / Natal Chart / December 11, 2002, Wed / 8:48:33 am PST +8:00 Reno, NV / 39°N31'47" 119°W48'46" / Geocentric / Tropical / Placidus / Mean Node*

OUTER WHEEL=*Wind / Natal Chart / December 14, 2002, Sat / 3:00 pm PST +8:00 Reno, NV / 39°N31'47" 119°W48'46" / Geocentric / Tropical / Placidus / Mean Node*

Three days after the lunar phase, Mercury had advanced to 9 Capricorn, where it was conjunct the lunar phase Ascendant and, most important, square ingress retrograde Mercury, making a strong connection between the two charts and activating the potential of the ingress Mercury. At the same time, transiting Mars and Venus were sextile the lunar phase Ascendant, another indicator of southerly winds. Another high wind aspect, Mars parallel Uranus, was within a 1-degree orb at the lunar phase, and exact on December 14.

This example illustrates another factor to look for when comparing ingress and lunar phase charts. Notice that on the date of the wind event, Uranus had returned to its position at the ingress (and also repeating the Mercury-Uranus aspect at the ingress), having retrograded and turned direct in the interim. This phenomenon, along with transiting planets triggering an aspect in the ingress chart, is often seen at the time of weather events.

The wind abated at about 6:00 pm, when the Moon completed its transit of Aries and entered Taurus. The Moon's sign change is often an excellent timer of the beginning or ending of a weather event.

These charts illustrate another principle in weather forecasting. Notice Neptune conjunct the ingress Midheaven at 9 Aquarius. When the chart is turned counterclockwise (representing a location west of Reno), Neptune is exactly conjunct the Midheaven, and lunar phase and transiting Venus and Mars in Scorpio are square that planet and point. At the time that Reno was experiencing high winds, the California and Oregon coasts were being drenched with rain accompanied by high winds. Some areas received more than nine inches in twenty-four hours. So even if you looked only at the charts for Reno, you could easily predict significant precipitation to the west.

Salt Lake City, Utah, April 3, 1964

Sixty mile per hour winds blew in Salt Lake City, Utah, on April 3, 1964, pushing trucks and trailers off highways. Pilots flying in the vicinity reported unusual turbulence and strong wind over the mountains that surround the city.

The potential for a windy season is apparent in the spring ingress chart (inner wheel of Wind 2), with Mercury square the meridian and Uranus sextile/trine the IC/Midheaven. Two aspects to Uranus indicate wind potential as well: Mercury inconjunct Uranus (gusts) and Jupiter sesquisquare Uranus (high wind). Pluto's wind-making potential was expanded by its high declination (20N01) and the sesquisquare from Jupiter. The Moon, parallel the IC, was square Mars (wind) and trine Saturn (low pressure), which was semisquare the Ascendant. The lunar eclipse prior to the ingress (December 30, 1963) occurred at 8 Cancer, conjunct the ingress IC. These aspects indicated the potential for high wind, but an appropriate lunar phase chart and aspects by the transiting planets were also necessary to confirm that a weather event would actually materialize.

Mercury's placement in the lunar phase chart of March 27, 1964 (middle wheel of Wind 2), doubled the effects of the ingress Mercury. In the lunar phase chart, Mercury was conjunct the Descendant and square the meridian, and formed two aspects to Uranus: sesquisquare and parallel (high wind, gusts). Jupiter was also square the meridian and was parallel Mercury and Uranus (high wind), which in turn was semisquare the Ascendant and sesquisquare the IC. Uranus also formed an inconjunct with the Sun (gusty winds). The Mercury-Uranus-Jupiter combination brought a high-pressure system to the area with the low pressure necessary to produce tight pressure gradients and the increased wind velocity associated with Mars sesquisquare Neptune, an aspect that indicates a rapidly falling barometer. At the solar eclipse of January 14, 1964, the Sun and Moon were at 23 Capricorn 43, conjunct the lunar phase IC.

In addition to the presence of the Mars-Saturn semisextile (wind) in the lunar phase chart, this aspect also connected the two charts because Mars had advanced far enough by the date of the lunar phase to aspect ingress Saturn, which was semisquare the ingress Ascendant. Lunar phase Jupiter was sextile the ingress Moon, activating the lunar aspects in the ingress chart.

However, the major planetary configuration between the lunar phase and ingress charts that indicated ideal conditions for a high-wind event was the lunar phase Sun conjunct the ingress Mercury, with both square the ingress meridian. To further connect the two charts, lunar phase Saturn was in a closer semisquare to the ingress Ascendant than it was at the ingress.

On April 3, 1964 (outer wheel of Wind 2), the transiting Sun at 14 Aries triggered Saturn (low pressure) in all three charts through a semisquare. Jupiter (high pressure) at 28 Aries was sesquisquare Pluto (high wind) in high declination in all three charts and parallel Uranus (high pressure and high wind). Mars was at 0 degrees, transitioning from southern to northern declination, which often occurs at a weather event because it intensifies the effects of a planet. In this case, the effect was compounded because Mars was parallel the ingress Sun (unsettled atmosphere, windy). Transiting Mercury was contraparallel ingress and lunar phase Saturn (wind, low pressure).

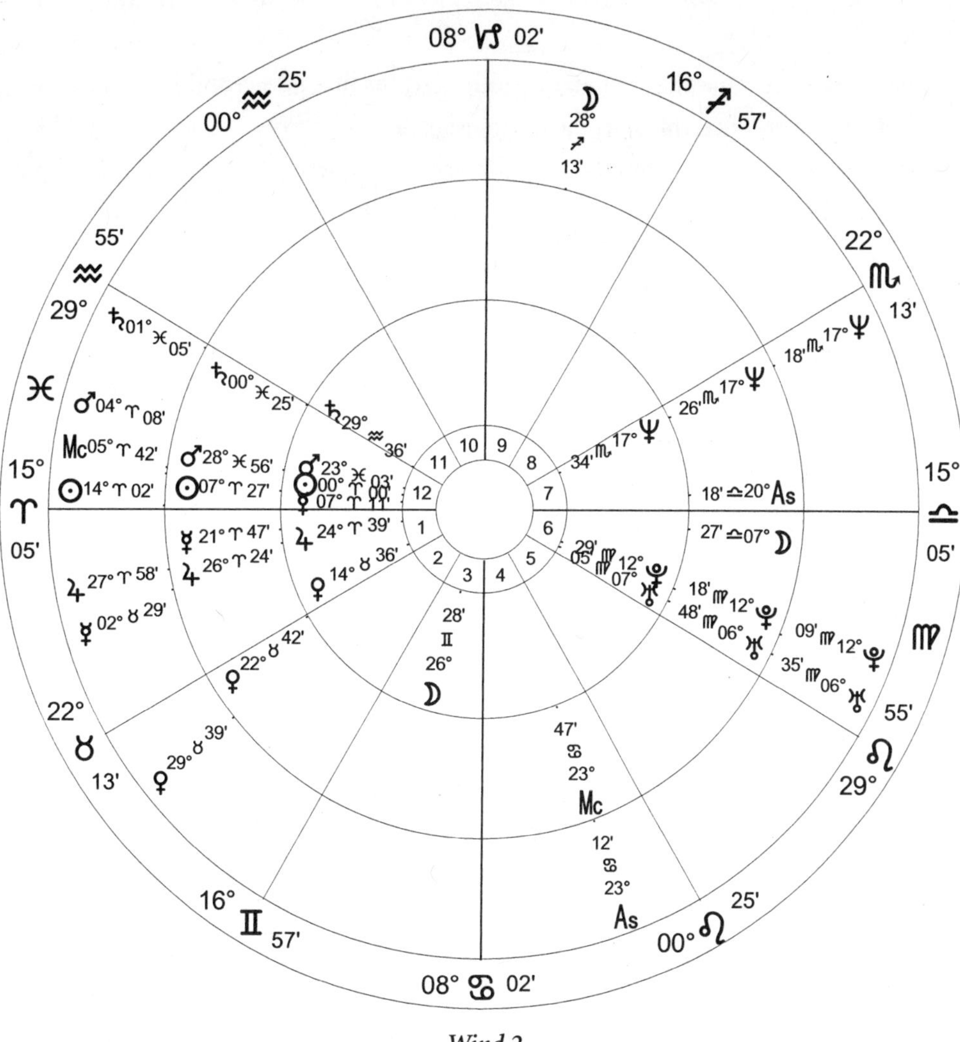

Wind 2

INNER WHEEL=*Spring Ingress / Natal Chart / March 20, 1964, Fri / 7:10 am MST +7:00 Salt Lake City, UT / 40°N45'39" 111°W53'25" / Geocentric / Tropical / Placidus / Mean Node*

MIDDLE WHEEL=*Lunar Phase / Natal Chart / March 27, 1964, Fri / 7:48 pm MST +7:00 Salt Lake City, UT / 40°N45'39" 111°W53'25" / Geocentric / Tropical / Placidus / Mean Node*

OUTER WHEEL=*Wind / Natal Chart / April 3, 1964, Fri / 12:00 pm MST +7:00 Salt Lake City, UT / 40°N45'39" 111°W53'25" / Geocentric / Tropical / Placidus / Mean Node*

Contacts from transiting planets to ingress chart angles (longitude and latitude of Salt Lake City) were numerous: transiting Sun at 14 Aries conjunct and parallel ingress Ascendant, transiting Saturn at 1 Pisces semisquare ingress Ascendant, and the Moon contraparallel ingress IC. The lunar phase IC was in parallel aspect with the transiting Moon.

The significant transiting planetary configurations that day were the same as those to the lunar phase and ingress charts with one significant exception: Mercury at 2 Taurus semisextile Mars at 4 Aries. On its own, that aspect excites westerly winds, but in combination with the Mercury-Uranus sesquisquare and parallel in the lunar phase chart, wind velocity reached gale strength (55 to 63 mph).

Southern California, November 30, 2011

A well-known weather phenomenon in Southern California is the Santa Ana winds. These strong downslope winds blow through mountain passes in their descent from the Mojave Desert when a high-pressure system, along with its clockwise air flow, dominates to the east. The strongest Santa Ana winds occur when there is a low-pressure system just to the north. The relatively short distance between the two systems typically increases wind velocity to 40 to 50 mph, with gusts in some places of as much as 100 mph. The descending cold, dry air of the high-pressure system is heated in its descent and arrives as a dry, hot wind, warming 5 degrees for every 1,000 feet of altitude it drops. These high-velocity dry winds often fuel intense fires. The charts shown (for San Bernardino) are representative of those throughout the affected area.

The possibility of destructive Santa Ana winds was apparent in the autumn ingress chart of 2011 (inner wheel of Wind 3). In fact, a National Weather Service employee said the area had not seen such a strong Santa Ana in more than a decade. More than a hundred trees were downed by the wind, tractor-trailers toppled, and more than 340,000 people were without power. The highest wind gust recorded in the mountains was more than 100 mph, and one gust hit 73 mph in the Los Angeles Basin.

The most striking wind-related alignment in the ingress chart was Mars (wind) at 2 Leo conjunct the Ascendant. Dryness is reflected in all the signs on the angles of this chart—Aries, Leo, Libra, and Aquarius—as well as the Sun sextile Mars. Mars was inconjunct Pluto (wind) and trine Uranus (high pressure).

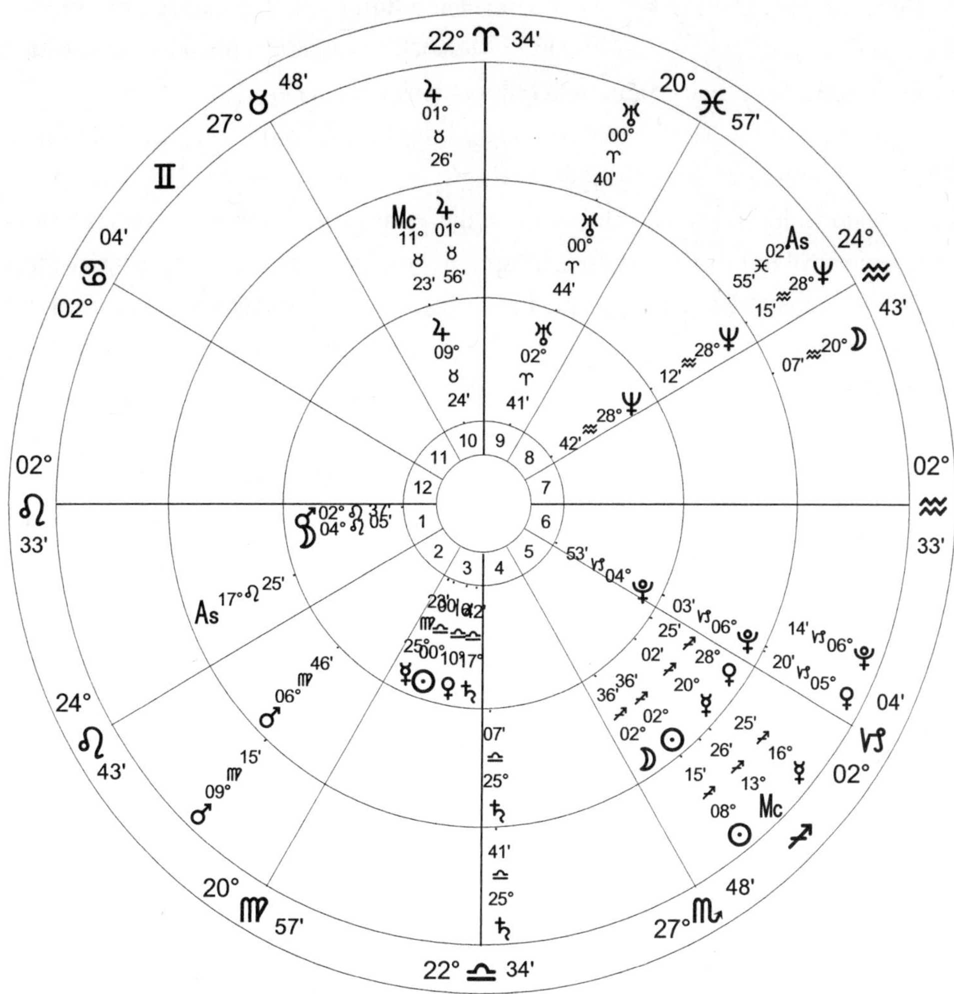

Wind 3

Inner Wheel=*Autumn Ingress / Natal Chart / September 23, 2011, Fri / 2:05:14 am PDT +7:00
San Bernardino, CA / 34°N07'17'' 117°W18'08'' / Geocentric / Tropical / Placidus / Mean Node*

Middle Wheel=*Lunar Phase / Natal Chart / November 24, 2011, Thu / 10:09:40 pm PST +8:00
San Bernardino, CA / 34°N07'17'' 117°W18'08'' / Geocentric / Tropical / Placidus / Mean Node*

Outer Wheel=*Santa Ana Wind / Natal Chart / November 30, 2011, Wed / 12:00 pm PST +8:00
San Bernardino, CA / 34°N07'17'' 117°W18'08''/ Geocentric / Tropical / Placidus / Mean Node*

Notice that Saturn (low pressure) is north and Jupiter (high pressure) is south. Combined with Mars trine Uranus, linking another high-pressure planet to that area, the setup was ideal for a major Santa Ana event during the season.

By the date of the lunar phase about two months after the ingress (middle wheel of Wind 2), Saturn was closer to the ingress IC and Jupiter had retrograded to 1 Taurus, forming a square with ingress Mars. In addition, the lunar phase Midheaven was conjunct ingress Jupiter. All of this reflects the atmospheric conditions necessary for and in effect at the time of a major Santa Ana event. There was a high-pressure system (Jupiter) to the southeast and a low-pressure system (Saturn) to the north. This produced the tight pressure gradients that resulted in very high winds.

Adding more wind energy to the event was lunar phase Mercury making its retrograde station at 20 Sagittarius in an approaching trine to the lunar phase Ascendant. This aspect was well within orb by November 30 (outer wheel of Wind 3), because Mercury had retrograded to 16 Sagittarius. On the date of the event, Mars was trine the lunar phase Midheaven and semisquare lunar phase Saturn conjunct the ingress IC.

There were many connections that linked the lunar phase chart to the ingress chart, emphasizing low pressure, high pressure, and wind at the Southern California longitude and latitude. The lunar phase Sun-Moon conjunction was trine the ingress Ascendant and contraparallel ingress Mars, sextile ingress Sun, and semisquare ingress Saturn. Stationary retrograde Mercury was sextile ingress Saturn, lunar phase Jupiter was square the ingress Ascendant-Mars, lunar phase Mars was sesquisquare/semisquare the ingress meridian, and ingress Jupiter was conjunct the lunar phase Midheaven.

CHAPTER 5

Drought

People sometimes react to news of drought as if it were something unusual or a new weather phenomenon; in fact, it is a part of the normal climatology, just like other forms of weather. The Pilgrims experienced drought in 1621 and 1623, and fields of grain in New England caught fire in 1749, 1761, and 1762. There were many droughts in the 1800s, including one on the Great Plains from 1881 to 1887. In the twentieth century, farmers battled drought in the 1930s, 1950s, 1960s, 1970s, and 1980s in varying locations throughout North America, and in the twenty-first century, drought has been a concern in western, southeastern, and central parts of the United States.

Drought is a relative term that generally can be defined as abnormal dryness in a particular region. If, for example, Alabama were to receive only twenty-four inches of rain in a year, or half its normal amount, the state would be in a severe drought condition. However, if Nevada received the same twenty-four inches, the opposite would be true, as its norm is well below that.

So drought does not mean there is no precipitation, but rather a diminished supply. At the same time, temperatures are often well above average during drought periods, causing moisture to rapidly evaporate. Hydrologists measure drought using the drought severity index, which measures soil moisture, evaporation, and precipitation over a long-term period. Normal conditions prevail when the index is between -2 and

+2, moderate drought between -2 and -3, drought between -3 and -4, and severe drought below -4.

The primary planets of drought are Mars, Jupiter, and Pluto, especially when in aspect with each other. If all three are active in an ingress or lunar phase chart, excessively high temperatures and dryness can be expected. The same is true when these planets are parallel, conjunct, or square the horizon or meridian.

High temperatures result when Mars is in close conjunction with the Sun, and any aspect between Mars and the Sun, Moon, Venus, or Mercury usually indicates excessive heat. This is even more true when these planets are in Leo, and sometimes in Aries. Saturn and Uranus, especially when in Gemini, Leo, Virgo, or Aquarius, can forecast dryness in colder seasons. Uranus conjunct the Midheaven indicates cold, dry air dipping far to the south, and Uranus and Jupiter reflect dry high-pressure systems. Planetary aspects denoting windy conditions are important as well, because wind pulls moisture from the soil, adding to the dryness.

Drought is not a singular event that spans a week or a month. It is a trend that affects a season, several seasons, a year, a series of years, or even a decade. Forecasting drought thus requires a succession of charts that indicate dryness, wind, and above-average temperatures.

Colorado, 2002

Colorado was in a severe drought in 2002, with a drought severity index of -8.74 throughout much of the year. According to the National Oceanic and Atmospheric Administration (NOAA), an index that low had not been seen in the region since 1685. That conclusion was based on tree ring analysis. Mandatory limits regarding water use were imposed in Denver for the first time in twenty-one years. High temperatures, rapid evaporation, and snowpack that was only 53 percent of average contributed to the severe drought.

April was very warm and dry, and only one significant storm occurred in May. There was some rain in June, but little soaked into the dry soil. Reservoirs were depleted, stream flows were low, heat was intense, and fires raged. July was very hot and dry, and there were more fires and extreme heat in August. Some precipitation finally arrived in September.

The three ingress charts covering these months and the three months prior all show a predominance of dryness and heat.

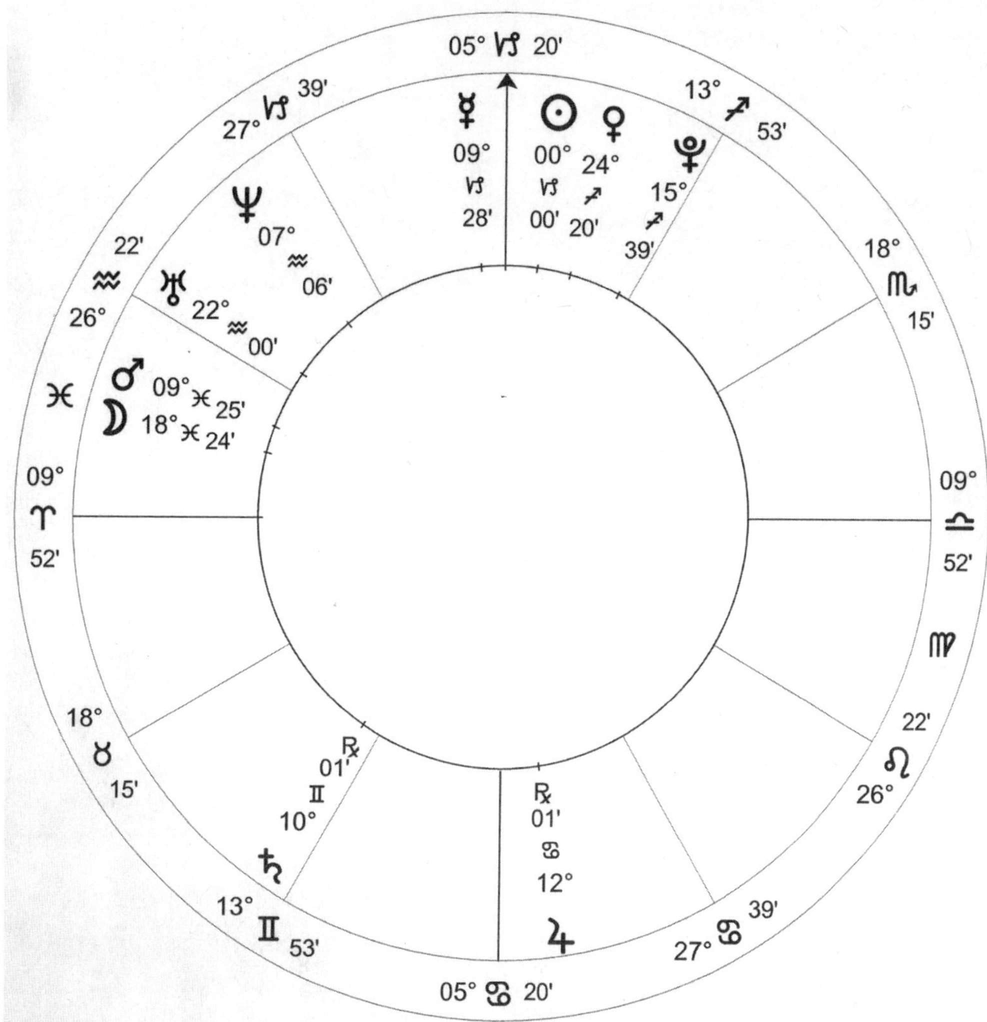

Drought 1: Winter Ingress / Natal Chart / December 21, 2001, Fri / 12:21:30 pm MST +7:00
Denver, CO / 39°N44'21" 104°W59'03" / Geocentric / Tropical / Placidus / Mean Node

The winter ingress chart had Mars trine Jupiter, with Jupiter inconjunct Pluto, square the horizon, and opposition Mercury (Drought 1). Mercury was also square the horizon and sextile Mars, which was semisextile the Ascendant. Uranus was semisquare/sesquisquare the horizon and meridian, with Venus in Sagittarius (fire sign) sextile Uranus and contraparallel Jupiter. The Mercury influence indicates wind that aggravated rapid evaporation and compounded the heat and dryness factors of the other planets.

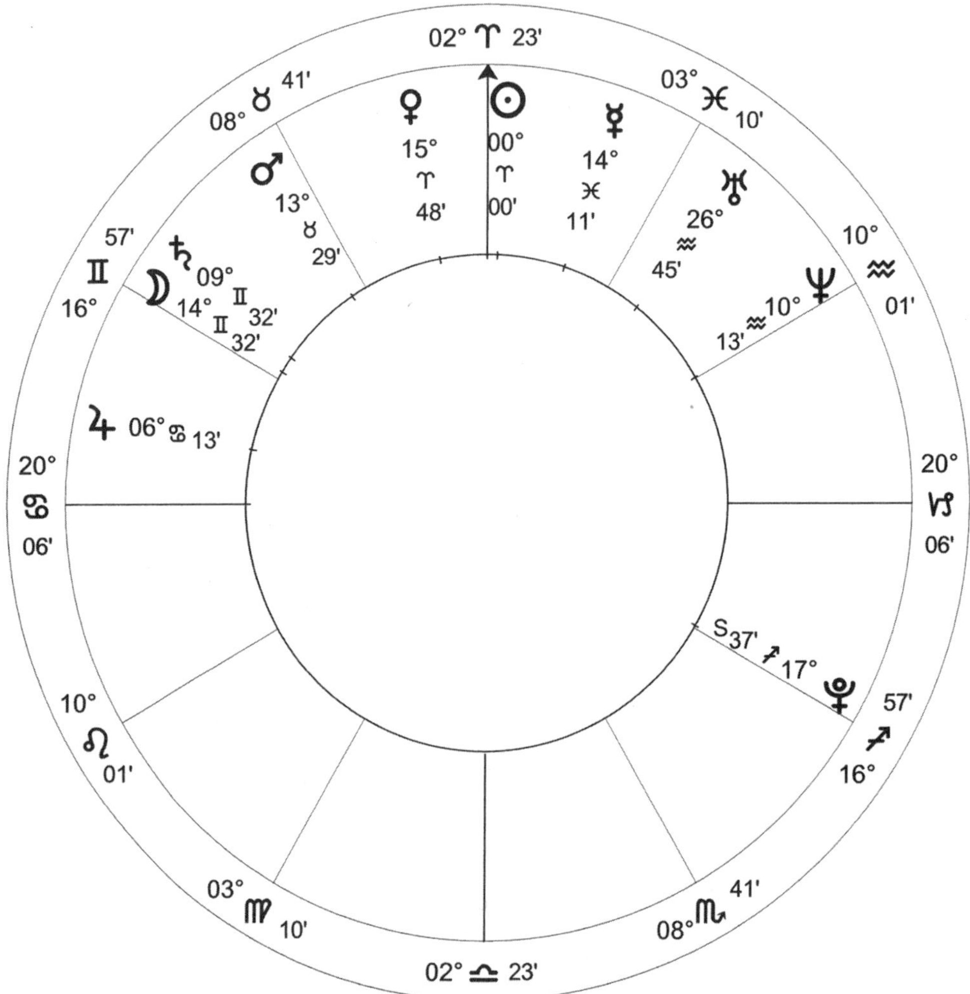

Drought 2: Spring Ingress / Natal Chart / March 20, 2002, Wed / 12:16:09 pm MST +7:00
Denver, CO / 39°N44'21" 104°W59'03" / Geocentric / Tropical / Placidus / Mean Node

At the spring ingress, the Sun in Aries, a fire sign, was conjunct the Midheaven, Jupiter was square the meridian, and Mercury was square Pluto. Venus, also in Aries (emphasizing heat, not moisture), was square the horizon and trine Pluto, which was parallel Uranus and inconjunct/semisextile the horizon (Drought 2).

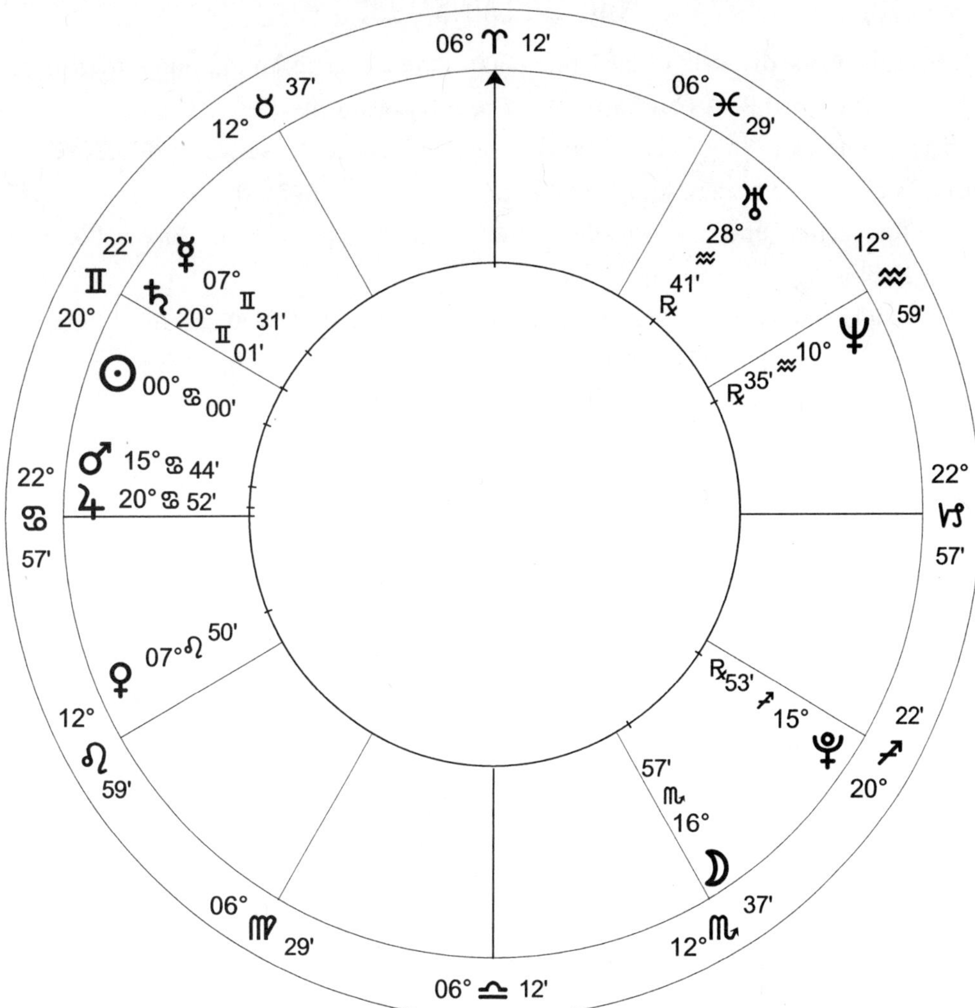

Drought 3: Summer Ingress / Natal Chart / June 21, 2002, Fri / 7:24:25 am MDT +6:00
Denver, CO / 39°N44'21" 104°W59'03" / Geocentric / Tropical / Placidus / Mean Node

Indicators of heat and dryness increased with the summer ingress (Drought 3). Mars was conjunct Jupiter conjunct the Ascendant, Mercury was semisquare/sesquisquare the horizon and sextile/trine the meridian, and Venus in Leo was trine/sextile the meridian and sextile Mercury. In addition to the Uranus-Pluto parallel in effect at the spring ingress, Venus and Mars were parallel Jupiter, with Mars adding intensity through its conjunction with Jupiter.

GREAT PLAINS, 1930S

One of the worst disasters in the United States was a decade-long drought that affected not only the Great Plains but most of the country at various times. The most seriously affected states, known collectively as the Dust Bowl, were Kansas, Colorado, New Mexico, Oklahoma, and Texas. At its worst in the winter of 1935–36, the Dust Bowl included 50 million acres. Other severely affected states were Montana, North Dakota, South Dakota, Nebraska, and Wyoming.

Drought first visited the Great Plains in the latter half of 1930 and early 1931. Welcome rains were present later in 1931, but dry weather returned in 1932 and continued year after year, in some areas until 1941. The summer of 1932 was a scorcher, only to be topped by 1934 and 1936. The area affected by drought began to shrink in 1937 and was down to 9.5 million acres by early 1939. However, extreme drought returned in 1939, to be followed finally by adequate precipitation in 1940 and early 1941.

Because the affected area was so widespread, a longitude and latitude in western Kansas, the center of the Dust Bowl, is used here to illustrate the planetary configurations. It is apparent from the charts that this location was severely affected, but it's important to remember that dryness was widespread. Because it is beyond the scope of this book to review all forty ingress charts for the decade, only a selected few are included. Especially notable factors in the others are mentioned, however.

In August 1934, the drought severity index for western Kansas was -5.96; August 1936 and October 1939 were a close second with -5.55. Between August 1932 and October 1940, the index was consistently below normal. During the worst years, the area received about eleven inches of annual rainfall, less than 60 percent of normal.

The September 23, 1930, ingress, which was the beginning of what would come to be known as the Dirty Thirties, shows the onset of dry conditions (Drought 4). Mars was conjunct and parallel Jupiter, and both planets formed an inconjunct and a contraparallel with the Ascendant. Both planets were very high in northern declination, and Jupiter also formed a conjunction and parallel with Pluto. All three major planetary configurations associated with drought were present, and all aspected the horizon.

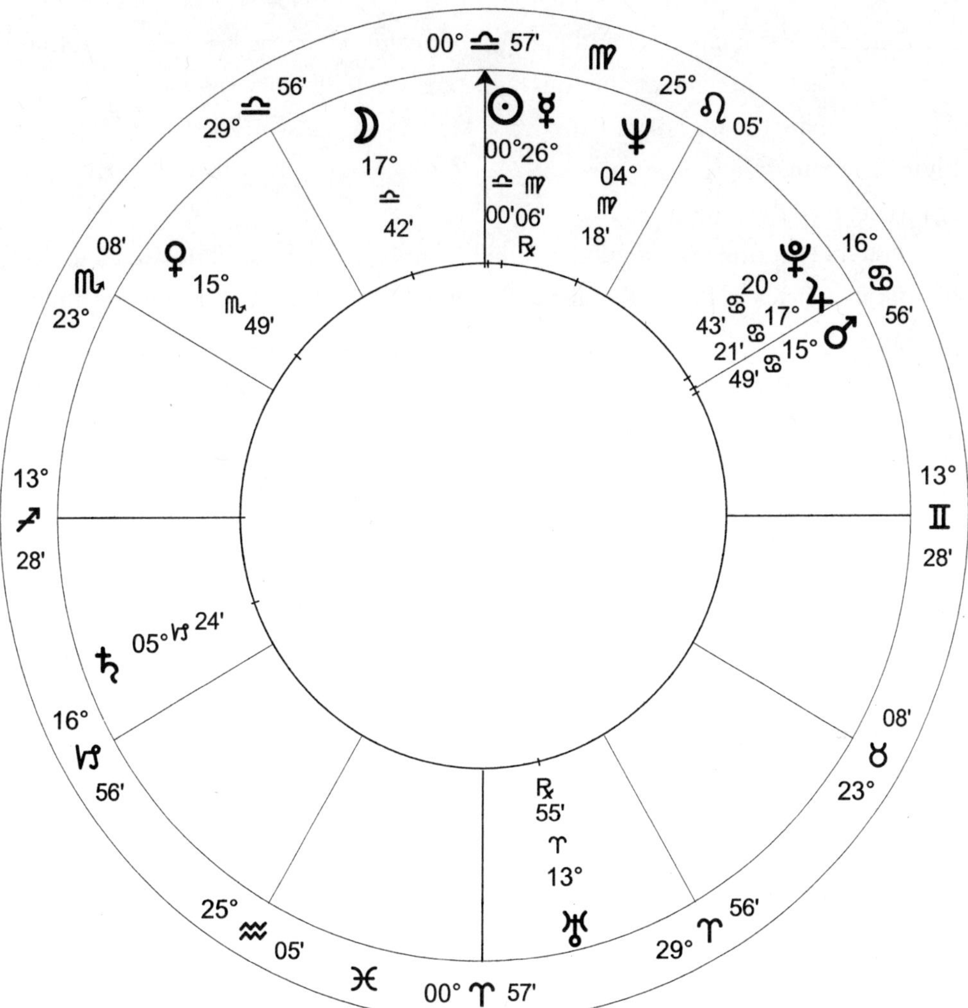

Drought 4: Autumn Ingress / Natal Chart / September 23, 1930, Tue / 12:36 pm CST +6:00
Great Plains / 38°N00' 100°W00' / Geocentric / Tropical / Placidus / Mean Node

Other indicators of higher temperatures were Venus trine Mars, Mars semisquare Neptune, and the long-term aspect of Neptune semisquare Pluto. Windy conditions were promised with the Sun conjunct and parallel retrograde Mercury, and Mars and Jupiter square Uranus, which was trine the Ascendant, reflecting a high-pressure, low-moisture pattern.

Dry conditions continued throughout the winter of 1930–31. At the ingress, Jupiter and Pluto were still conjunct and parallel, opposing Mercury. Jupiter and Pluto were contraparallel the Ascendant and square the meridian. Mercury was parallel the Ascendant and square the meridian. Retrograde Mars formed a semisextile with Jupiter and Pluto. The Sun, in a sesquisquare with Mars, was parallel the Ascendant, and Uranus again occupied the lower meridian.

At the 1931 spring ingress, Jupiter and Pluto were parallel and Pluto was conjunct the IC. Mars, Jupiter, and Pluto were all parallel, and Mars was square the horizon and trine the Sun.

The heat and dryness abated somewhat later that year, but returned in the summer of 1932. At that ingress, Jupiter was conjunct and parallel the Ascendant and contraparallel the IC, trine Uranus, and semisextile Pluto, which was semisextile/inconjunct the horizon. Mars was semisquare Venus and semisextile the Sun. The Sun, Mercury, Venus, and Jupiter aspected the meridian. Uranus, in a square with Pluto, was trine/sextile the horizon.

By that winter's ingress, Jupiter formed a sextile with Pluto conjunct the Ascendant; Jupiter-Pluto is probably the single most indicative aspect of severe drought. Uranus was square Pluto and semisextile and contraparallel Mars. The sextile was still present in the spring 1933 ingress chart.

As the heat and dryness continued into the spring and summer of 1933, the worst was yet to come. That autumn, the first of massive, devastating dust storms occurred on November 13, 1933. Nicknamed "black blizzards," the storms were dirt-filled black clouds as much as 20,000 feet high that roared across the Great Plains, piling soil in drifts. Visibility was reduced to a few feet at times as the roiling mass masked the sun. The worst of the storms were so powerful that skies from the Great Plains to the East Coast were darkened, and soil settled on vessels in the Atlantic Ocean.

The September 23, 1933, ingress chart shows the continuing drought and potential for black blizzards (Drought 5). Mars was semisquare Jupiter and the Sun, which were conjunct and parallel. Uranus was trine the IC and inconjunct/semisextile the horizon, while Pluto formed an inconjunct/semisextile and contraparallel with the meridian and a sextile/trine with the horizon.

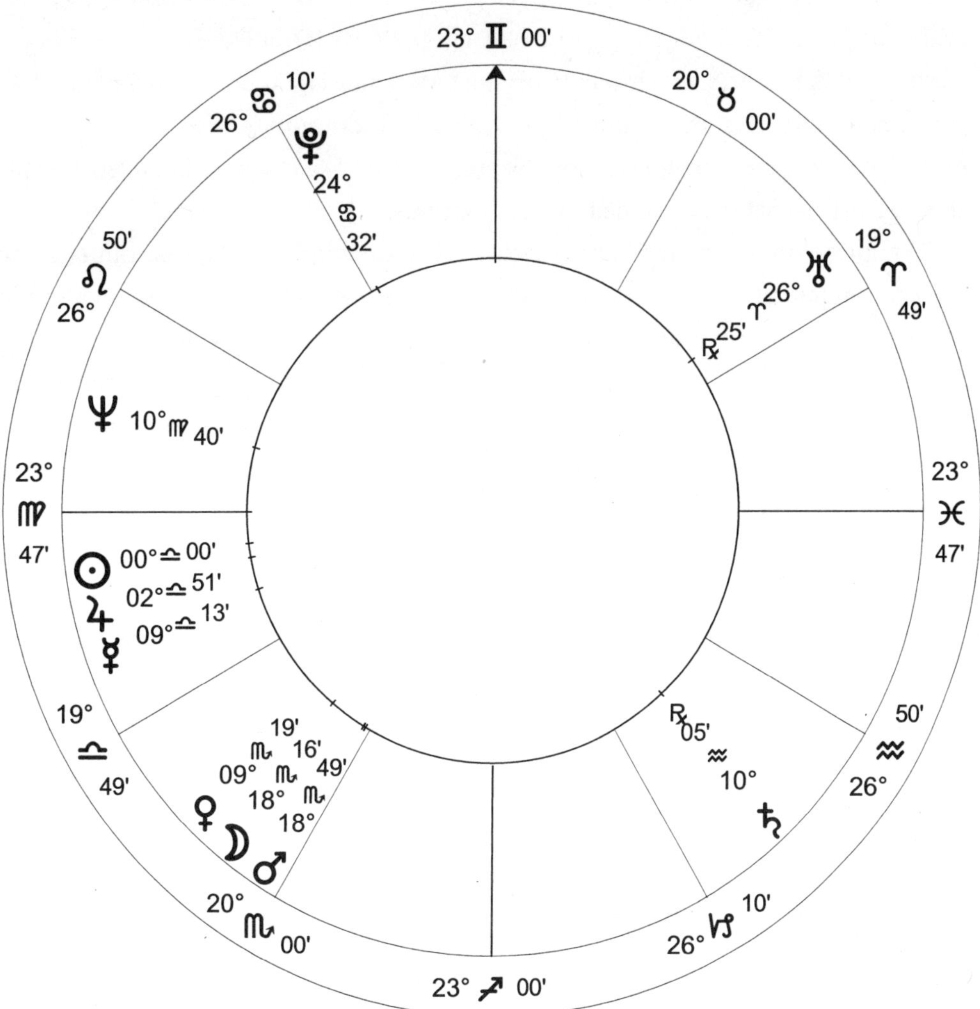

Drought 5: Autumn Ingress / Natal Chart / September 23, 1933, Sat / 6:02 am CST +6:00
Great Plains / 38°N00' 100°W00' / Geocentric / Tropical / Placidus / Mean Node

The Uranus-Pluto square, also an aspect of high wind, was still within orb, and Mars was parallel Saturn (windy storm), which was sesquisquare/semisquare the meridian and horizon. Mercury was contraparallel the Ascendant. The Uranus-Neptune sesquisquare (unusual weather), which had been within orb more often than not since the summer of 1930 because of the retrograde pattern of the planets, was nearly exact.

Mercury, the planet of wind, turned retrograde the day before the lunar phase of November 10, 1933, activating its negative retrograde effects (Drought 6). It was contraparallel Pluto (high wind), which was inconjunct Mars (gusts). Major storm potential was also apparent with Saturn conjunct the IC and in a semisquare with Mars, which was trine Uranus (dryness, high pressure). Neptune was still semisquare Pluto (high wind), and Jupiter trine Saturn expanded the storm energy.

There were many lunar phase aspects indicating wind and dryness but only two that showed potential for moisture: Uranus square Pluto (thunderstorms, but also high wind) and Neptune semisquare Pluto (low pressure, but also high wind).

Every planetary aspect was connected with the angles. In addition to Saturn conjunct the IC, Mercury was semisextile/inconjunct the horizon, Mars was semisquare/sesquisquare both the horizon and meridian, Jupiter was trine/sextile the meridian, and Neptune was inconjunct/semisextile the meridian.

There were significant aspects between the lunar phase and ingress charts, nearly all of them indicative of wind. Retrograde lunar phase Mercury was sextile ingress Jupiter at 0 degrees south (northerly wind) and contraparallel ingress Pluto (high wind). Lunar phase Mars was inconjunct ingress Pluto (wind) and thus also configured with these other Pluto aspects: lunar phase Uranus square ingress Pluto, and Pluto returning to its ingress position (high wind). Lunar phase Saturn was parallel ingress Mars (windy storm), and it also had returned to its position at the ingress.

Angular contacts between the lunar phase and ingress charts were numerous: lunar phase Mars was conjunct the ingress IC and square the ingress horizon, the lunar phase Ascendant was semisquare ingress Ascendant and sesquisquare the IC, and the lunar phase IC was square ingress Venus, semisquare the IC, and sesquisquare the Ascendant.

On November 13, key planetary configurations in the ingress chart were activated. Transiting retrograde Mercury at 2 Sagittarius was parallel Pluto and sextile Jupiter. Transiting Mars at 26 Sagittarius dominated with its trine to Uranus, inconjunct to Pluto, and semisquares to Venus and Saturn, and was conjunct the IC. Mars and Saturn were triggered by the transiting Sun in parallel to them, and the Moon in Virgo was conjunct the Ascendant, square the meridian, and parallel Mercury, Jupiter, and the Sun. Jupiter was parallel Mercury, further increasing the wind.

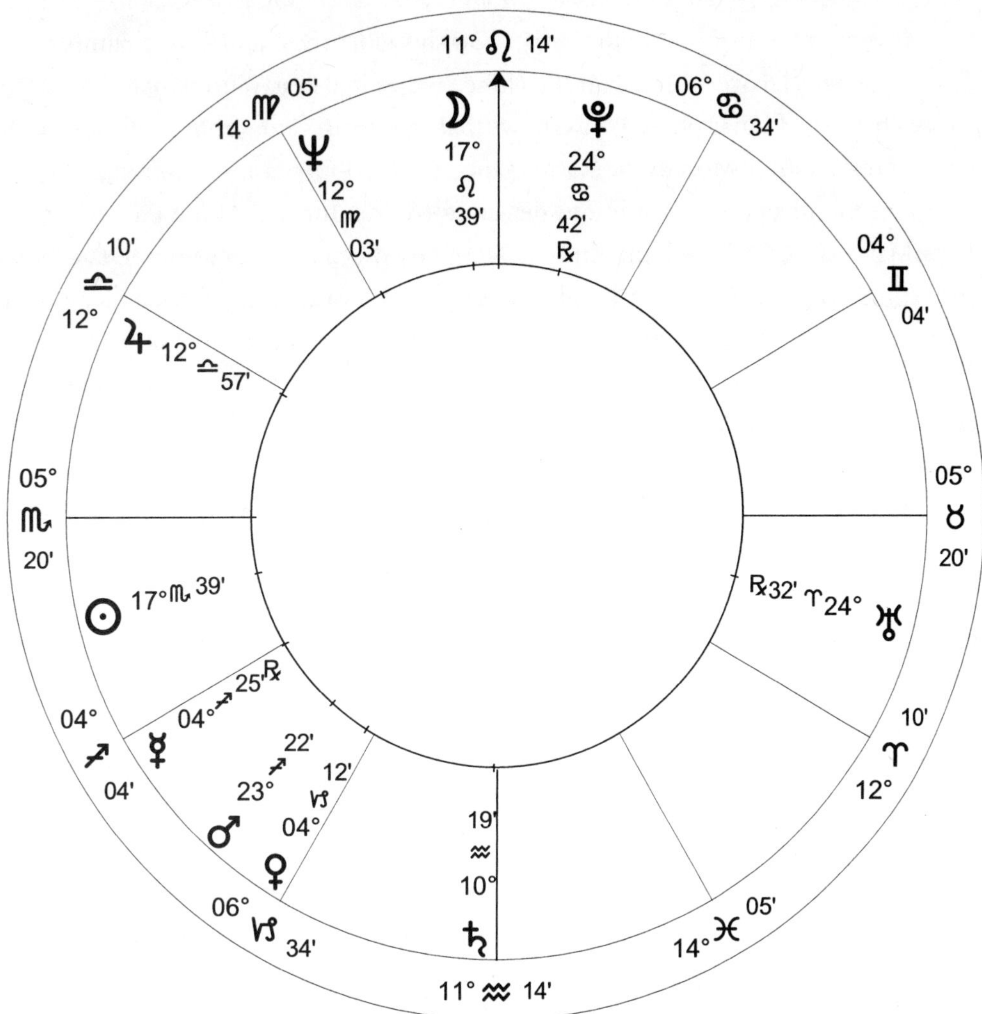

Drought 6: Lunar Phase / Natal Chart / November 10, 1933, Fri / 6:18 am CST +6:00
Great Plains / 38°N00' 100°W00' / Geocentric / Tropical / Placidus / Mean Node

The transiting Virgo Moon formed a square with lunar phase Mars and a semisquare/
sesquisquare with the horizon and meridian. Of equal significance was transiting Mars
semisquare/sesquisquare the lunar phase meridian, horizon, and Saturn. Retrograde Mercury was contraparallel Pluto, Uranus square Pluto, and the Sun parallel Saturn. Mars was
trine Uranus, inconjunct Pluto, and semisquare Saturn.

The Saturn involvement in the transits and lunar phase and ingress charts, and Saturn's aspects to the angles in both charts, show the widespread and lasting nature of the dust storm, which covered much of the United States in its eastward transit. The ninth-house position of lunar phase Pluto shows that the storm came from northwest of the point of observation. Most of the storms generated in Montana and Wyoming.

As the months passed, conditions deteriorated even further. At the 1933 winter ingress, Mars was opposition Pluto, and Jupiter opposed Uranus. Three months later at the spring ingress (March 21, 1934), Jupiter was square Pluto and sextile/trine the horizon. Mars was just past a conjunction with the Sun, and Uranus was square Pluto.

When summer 1934 arrived on June 21, Mercury was conjunct Pluto, Uranus was conjunct the IC, and Mars and Jupiter formed a trine. Mars was square Neptune and parallel Pluto to add to the stifling heat. Mars, Jupiter, and Pluto all aspected the horizon.

In 1935, the year of forty dust storms, the spring ingress occurred on March 21 with Pluto sesquisquare retrograde Mars in the seventh house, which formed a semisextile with retrograde Jupiter. At the summer ingress, Pluto was conjunct the IC and Jupiter conjunct the Descendant. That autumn, Mars was contraparallel the IC square Saturn conjunct the Ascendant, and also contraparallel Pluto, which aspected both the ingress horizon and meridian.

The hottest summer of the decade, in 1936, was preceded by a hot and dry spring, with Mars trine Jupiter and square the horizon, Uranus inconjunct/semisextile the meridian, and Pluto inconjunct and contraparallel Jupiter, which was contraparallel the Ascendant.

At the summer 1936 ingress, Pluto was conjunct and square the meridian, and Uranus was square the horizon. Mars was semisquare Uranus and semisextile and parallel Pluto. The Sun and Venus, conjunct Mars, were semisextile and parallel Pluto. This sizzling combination resulted in temperatures as high as 121 F in North Dakota, 108 F in Iowa, 113 F in Kansas, and 112 F on the Eastern Seaboard.

Although temperatures cooled somewhat after the summer of 1936, the planetary aspects of heat and drought persisted in western Kansas until 1940, when some of the thousands of families who had fled to California began to return to rebuild the Great Plains.

Northeastern United States, 1961–66

The first indications of drought that affected the Northeast in the 1960s appeared in September of 1961. By the end of October, it was classified as a mild to moderate drought covering the region from northern Virginia to southern Vermont. New York City reported a water shortage with reservoirs at 30 percent of capacity.

In July and August of the next year, the drought had become moderate to severe, but precipitation over the winter months eased the situation by spring 1963. Dry weather returned in October, when it was very warm, to be followed by six months of wet weather.

It was dry again in May 1964, and by September, many areas were classified in the severe drought category. June, July, and August 1965 were the driest yet, with New York City receiving 37 percent of its normal rainfall (the average for the entire year was 50 percent). There was moderate precipitation in January and February 1966, but spring and summer were dry. The drought ended in October 1966, beginning with heavy rain the day of the autumn ingress.

The summer 1965 ingress chart reflects the dry conditions that summer (Drought 7). Jupiter was square the Uranus-Pluto conjunction, combining the planet of high pressure with a major drought aspect. Mars was square the Sun.

Windy conditions were indicated in part by Mercury semisextile Jupiter and sextile Uranus and Pluto. Saturn's opposition to Pluto and the Uranus-Pluto conjunction showed potential for windy afternoon thunderstorms after hot days. Saturn and Venus trine Neptune, Neptune sextile Pluto, and the Sun sesquisquare Neptune helped keep the atmosphere steamy and cloudy, with the potential for showers. Jupiter was square Saturn, normally a configuration of above-normal precipitation. So the question is why all these configurations didn't manifest as much-needed moisture.

A look at the planets that aspected the angles reveals the explanation. Only the Sun and Venus formed any aspect to the horizon or meridian. The Sun formed an inconjunct/semisextile with the meridian, and Venus was sesquisquare/semisquare the meridian and horizon. With both these planets in Cancer, the potential for moisture was there, but drought prevailed until a lunar phase triggered the ingress chart; then the city received a little rain.

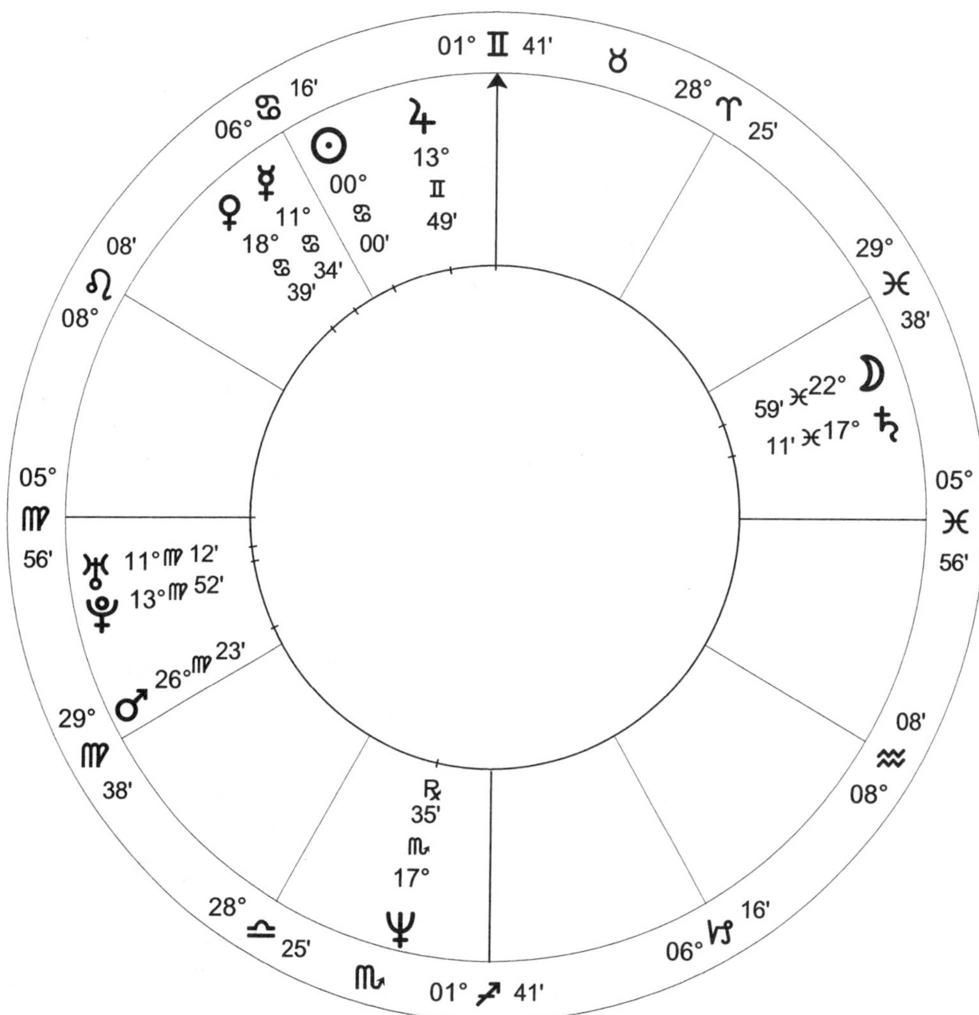

Drought 7: Summer Ingress / Natal Chart / June 21, 1965, Mon / 10:56 am EDT +4:00
New York, NY / 40°N42'51" 074°W00'23" / Geocentric / Tropical / Placidus / Mean Node

At the end of June, transiting Mars in Libra was square the ingress Sun, transiting Jupiter was parallel ingress Venus, and transiting Mercury was parallel Jupiter. Mars was at 0 degrees north declination, intensifying the heat and dryness.

In early July, transiting Venus and Mercury entered Leo (contraparallel the ingress IC) and formed a sextile with transiting Mars in Libra as transiting Jupiter formed a parallel with the ingress Sun and Venus. At the same time that transiting Mercury and

Venus were parallel transiting Pluto, they did the same with ingress Jupiter. At mid-month, transiting Mercury and Venus were sextile ingress Jupiter and contraparallel ingress Neptune, and the Sun was contraparallel the ingress IC. Transiting Mars in Libra was trine ingress Jupiter and square ingress Mercury and Venus at the end of the month as the Sun formed a parallel with ingress Pluto, a square and parallel with the ingress IC, and a conjunction with the ingress Ascendant.

Transiting Mercury, which turned retrograde on August 2 at 0 Virgo, quickly returned to Leo to semisextile the ingress Sun and square the ingress meridian, indicating even more heat and dryness. Transiting Mars in dry Libra was square ingress Venus, and transiting Venus in dry Virgo formed a conjunction with ingress Uranus and Pluto. Transiting Mercury and Venus were contraparallel transiting Mars and the ingress Ascendant, while the transiting Sun was parallel Pluto, and transiting Mars was contraparallel transiting Uranus.

By mid-August, transiting Venus was conjunct and Jupiter square ingress Mars, and both were contraparallel the ingress Ascendant. Transiting Uranus was parallel the ingress Ascendant at the end of the month, and transiting Venus was parallel ingress Mars. All month, transiting Jupiter was parallel the ingress Sun and Venus.

More than a year would pass before New Yorkers saw a normal amount of precipitation; the drought ended on September 23, 1966, when rain arrived. In the ingress chart, Jupiter did not aspect the horizon or meridian and formed only a few aspects: sextile the Sun, semisquare Venus, and trine Saturn. Mars was still square Pluto, but neither of these planets aspected the horizon or meridian.

Mercury was square the meridian and conjunct and parallel the Ascendant, along with Saturn parallel and Uranus contraparallel the Ascendant. These were key factors indicating rain for the drought-stricken area and were the only planets that aspected the ingress chart angles.

The aspects in the ingress chart were Mercury parallel Saturn (cloudy, falling barometer, precipitation, cooler) and contraparallel Uranus (overcast, rain, storm, cooler), and Mars square and contraparallel Neptune (low pressure, cloudburst). Cooler temperatures were indicated by Jupiter trine Saturn, Saturn contraparallel Uranus, and Mars semisextile Uranus. During the last week of September, transiting Venus and Mercury were parallel the ingress Ascendant (precipitation), and transiting Venus opposed ingress Saturn (heavy precipitation).

____

CHAPTER 6

____

Floods

Floods cause incredible destruction that can exceed that of hurricanes and tornadoes, affecting large areas and even entire regions. They also can have far-reaching effects that last for months, seasons, or years.

Floods generally occur in low-lying areas near water sources after extensive rain, or as a result of heavy rain combined with spring snowmelt. As the earth becomes saturated and unable to absorb any more moisture, the water table rises. When it reaches its peak at ground level, the water begins to flow into natural channels, rivers, and streams. As more and more rainfall is diverted to these bodies of water, they rise and overflow their banks, sometimes becoming virtual lakes. Standing water also builds up in low-lying areas when streams and rivers can accept no more water.

Flash floods are more common in the arid areas of the western United States. However, they can occur in any location that receives an unusually high volume of rain in a limited amount of time. In the western states, flash floods are the result of cloudbursts that produce heavy rain during a few hours' time. Because the earth is so hard and dry, it is unable to absorb the water as fast as it falls. Thus the rain is channeled into otherwise meandering streams and dry washes, where it races along over the earth or through mountainous valleys. Flash floods happen so quickly that people miles downstream, often under cloudless skies, receive little or no warning that a torrent of water is rushing their way.

A variety of atmospheric conditions cause floods. These range from long-term (an entire season or more) weather patterns to stationary low-pressure systems stalled in place for a day or two. Heavy precipitation during early spring months, when combined with unusually warm temperatures that hasten snow melt, can cause rivers and streams to rise and overflow their banks and levees.

Potential for flooding is indicated in ingress and lunation charts by a dominance of storm and precipitation planets such as Venus (precipitation), Saturn (low pressure, storms), and Neptune (low pressure, heavy precipitation). When aspects involving these planets are repeated over and over—ingress chart, followed by a succession of lunar phase charts—flood potential increases. Because storms are slower in developing and longer lived when Saturn is active, the combination of this planet with Venus and/or Neptune signals that precipitation is likely to extend over several days or even weeks. Strong aspects to the meridian are also necessary for flooding, such as Neptune conjunct, parallel, or square the meridian.

DYERSBURG, TENNESSEE, APRIL 2011

Flooding was widespread in the Mississippi River Valley in the spring of 2011 as a result of record levels of rain during consecutive weeks. Six states were severely affected, primarily in April and early May. About twenty people died as a result of the flooding, and damages were estimated to be more than $2 billion.

Dyersburg, Tennessee, experienced the worst flooding in the area, with damage and destruction to more than 600 homes and businesses. Many area rivers and creeks overflowed their banks, and the Forked Deer River, a tributary of the Mississippi, reached 47.8 feet on May 10. This was the second highest level ever recorded.

Rather than look at an individual rain event, we'll look at the ingress chart and then the lunar phase charts for the weeks of major storms and heavy rainfall that led up to the flooding of May 10. The most severe storms produced tornadoes in addition to heavy rain, and there was an increasingly strong presence of Neptune, the planet associated with flooding, as the weeks unfolded.

The March 20, 2011, spring ingress chart had Mercury and Jupiter opposition Saturn (storms), with all three planets aspecting the horizon and meridian and in declination with each other (Floods 1). Venus, which was parallel Neptune (heavy precipitation), was trine/sextile the meridian and semisextile/inconjunct the horizon, and also semisextile Mars, which was square the meridian and conjunct the Descendant (wind, warm air). Venus was also semisquare Pluto (heavy precipitation). The warm planets—Venus, Jupiter, and Neptune—indicated warm, moist air for thunderstorm formation, and Mercury and Saturn indicated cold air.

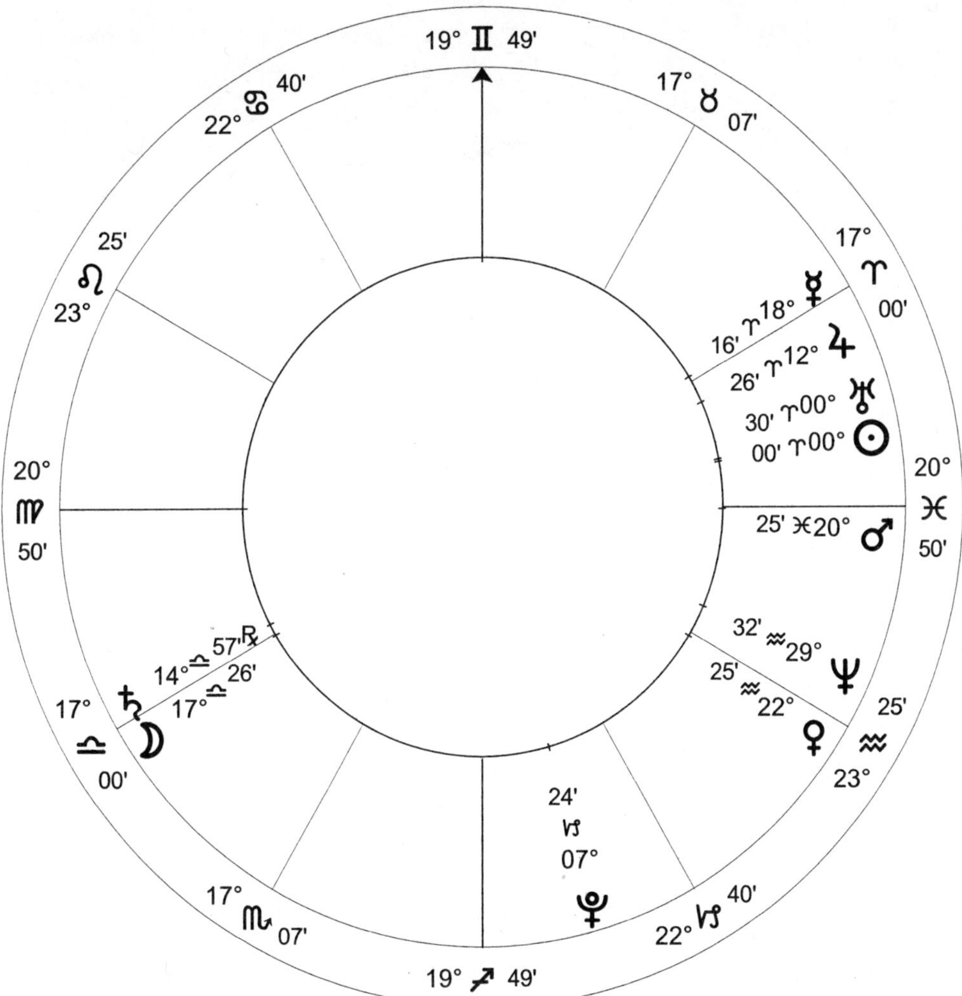

Floods 1: Spring Ingress / Natal Chart / March 20, 2011, Sun / 6:20:43 pm CDT +5:00
Dyersburg, TN / 36°N02'04" 089°W23'08" / Geocentric / Tropical / Placidus / Mean Node

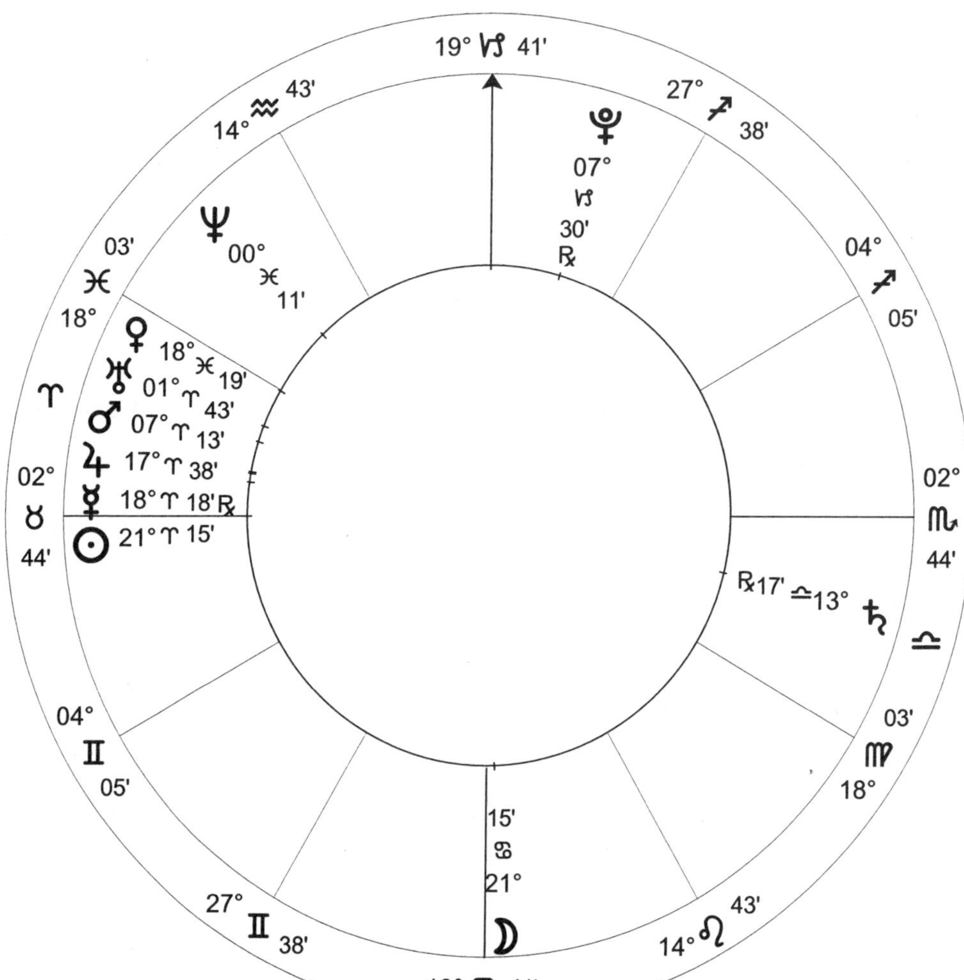

Floods 2: Lunar Phase / Natal Chart / April 11, 2011, Mon / 7:05:22 am CDT +5:00
Dyersburg, TN / 36°N02'04" 089°W23'08" / Geocentric / Tropical / Placidus / Mean Node

The lunar phase of April 11, 2011 (Floods 2), had Venus conjunct the ingress Descendant and square the ingress meridian, a classic placement for heavy precipitation. Lunar phase Mars was semisquare ingress Venus and transiting Neptune (heavy precipitation, tornadoes). Mercury was retrograde in this chart and conjunct lunar phase Jupiter (high wind) and ingress Mercury (high wind), semisquare Neptune (tornado, wind, precipitation), and square the lunar phase meridian. The lunar phase Sun was sextile ingress Venus (precipitation) and square the lunar phase meridian.

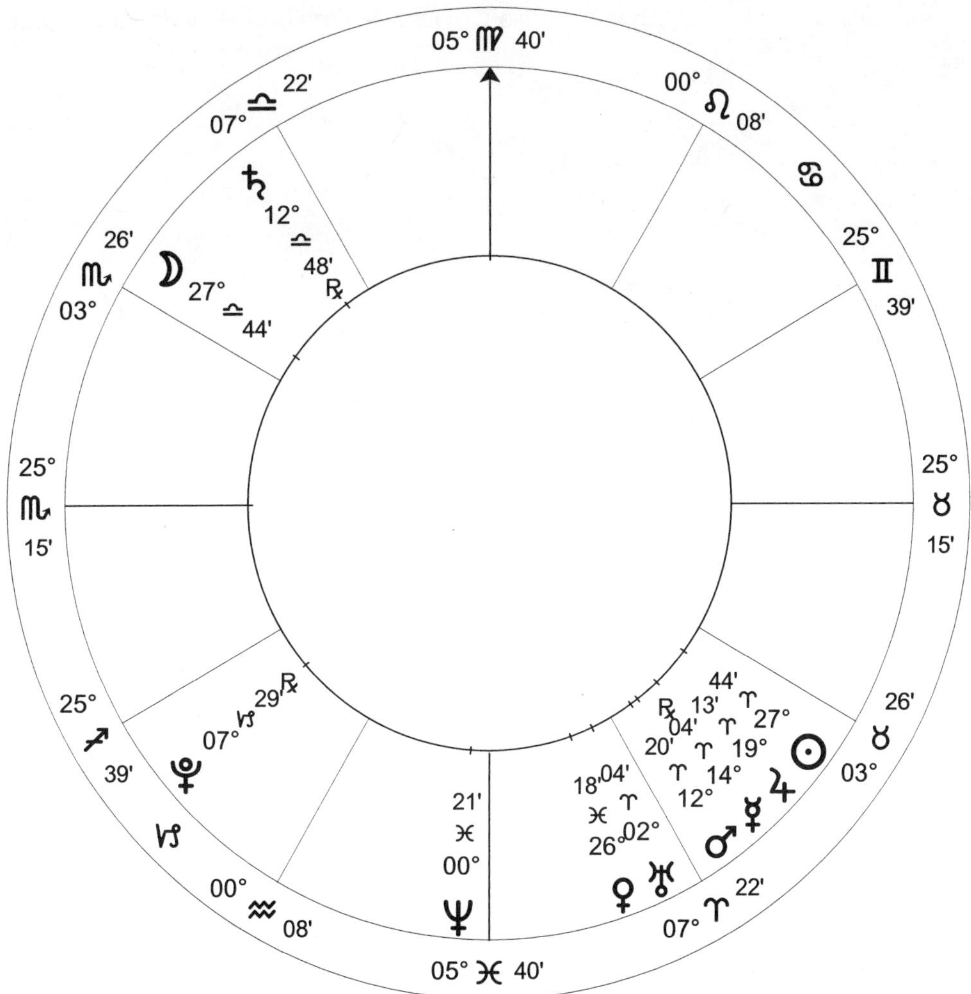

Floods 3: Lunar Phase / Natal Chart / April 17, 2011, Sun / 9:43:58 pm CDT +5:00
Dyersburg, TN / 36°N02'04" 089°W23'08" / Geocentric / Tropical / Placidus / Mean Node

Neptune had an even stronger presence at the April 17, 2011, lunar phase, with Neptune conjunct the lunar phase IC, a notable aspect of heavy precipitation and flooding (Floods 3). Aspects to Neptune were Venus semisextile, Mercury semisquare, and Sun sextile (heavy precipitation, tornadoes). Mars was conjunct Jupiter (thunderstorms), which was sesquisquare/semisquare the lunar phase meridian. The stormy combination of Mars opposition Saturn aspected the horizon. Indicating even more precipitation in Dyersburg was ingress

Venus (precipitation) sesquisquare/semisquare the ingress meridian and square the lunar phase Ascendant. Pluto, which was semisquare ingress Venus, was trine/sextile the lunar phase meridian.

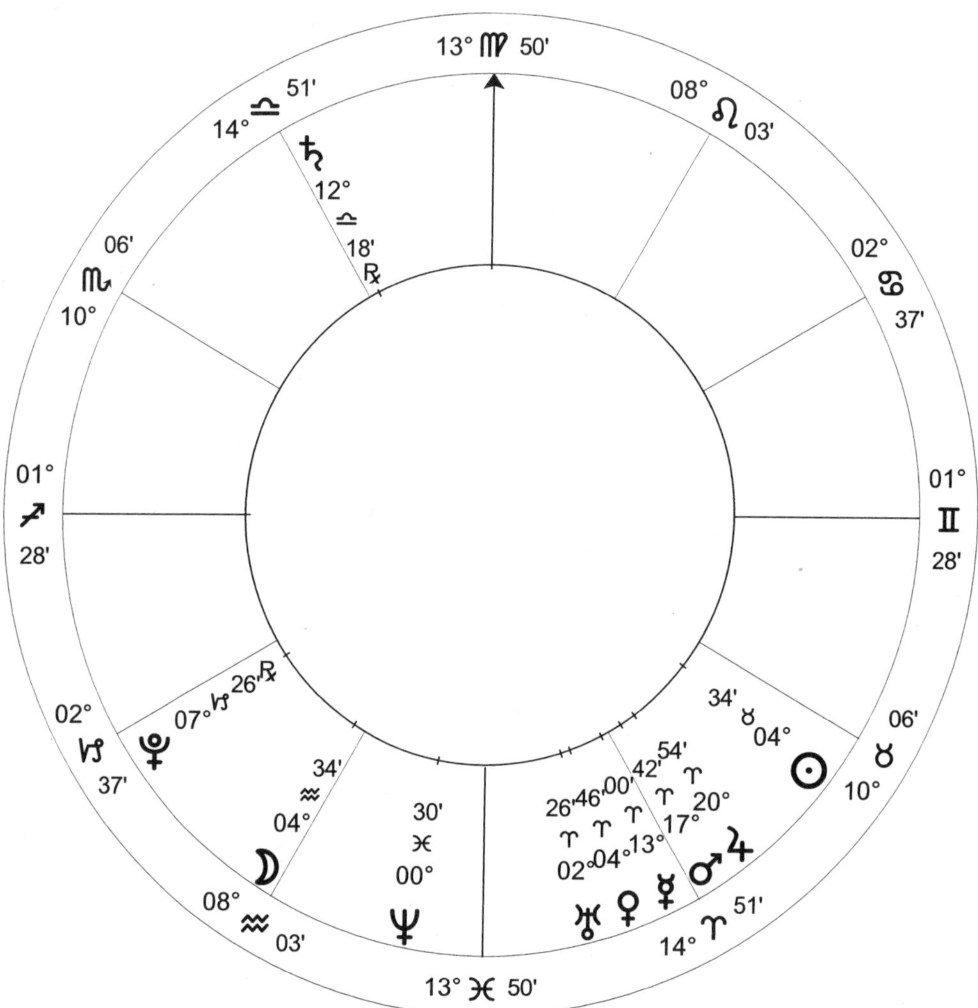

Floods 4: Lunar Phase / Natal Chart / April 24, 2011, Sun / 9:46:52 pm CDT +5:00
Dyersburg, TN / 36°N02'04" 089°W23'08" / Geocentric / Tropical / Placidus / Mean Node

Neptune was again strong at the April 24, 2011, lunar phase (Floods 4), where it was square the lunar phase horizon and semisquare Mercury (storm, tornado), which was in-conjunct/semisextile the lunar phase meridian and conjunct the ingress Jupiter (moist, southerly air). Mars was conjunct ingress Mercury (storm) and trine/sextile the ingress meridian. As the week unfolded, Venus formed a semisquare with ingress Venus and a square with Pluto (heavy precipitation).

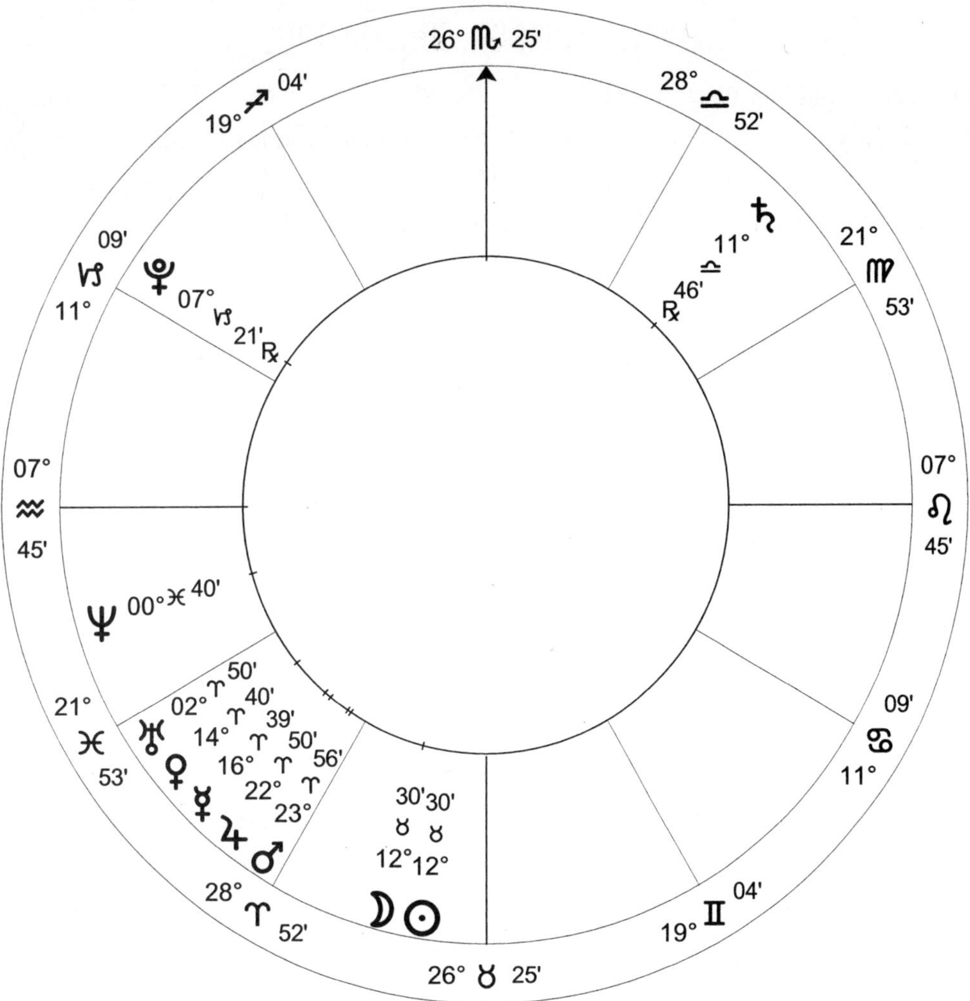

Floods 5: Lunar Phase / Natal Chart / May 3, 2011, Tue / 1:50:41 am CDT +5:00
Dyersburg, TN / 36°N02'04" 089°W23'08" / Geocentric / Tropical / Placidus / Mean Node

May 3, 2011, was the final lunar phase (Floods 5) before the May 10 flooding (also the date of the next lunar phase). Mercury was opposition ingress Saturn (storm, heavy precipitation), Mars was conjunct Jupiter (thunderstorms), Venus was opposition ingress

Saturn (heavy precipitation), and the lunar phase horizon was semisquare/sesquisquare the ingress horizon. The final aspect that indicated overflowing rivers and creeks was Neptune square the meridian.

HEPPNER, OREGON, JUNE 14, 1903

About 250 residents of Heppner, Oregon, lost their lives when a flash flood roared through the town of 1,500. Damaging and deadly, a flash flood is more limited in scope than other floods, arising very quickly when torrential rains rapidly fill small streams or rivers; this is in contrast to a series of storms over days or weeks that gradually fill rivers to overflowing. The Heppner flood is an extreme example of a flash flood.

A cloudburst erupted over Balm Fork Canyon, south of Heppner, and the water raced down the narrow canyon, picking up haystacks, chicken pens, and livestock that then piled up behind a laundry building that straddled the canyon near the edge of town. When this makeshift dam then failed, a wall of water from fifteen to fifty feet high rushed through the town, destroying about a third of Heppner's structures. Although the time of the flood varies somewhat in historical accounts, it occurred about 5:15 pm, shortly after the thunderstorm began.

The ingress chart (inner wheel of Floods 6) offers little to indicate the potential for a flash flood, at least from the perspective of typical precipitation and flooding aspects. What the chart does show is a tendency for dryness, which had indeed been the case that spring. This made the area even more susceptible to a flash flood because it was difficult for the dry, hard earth to absorb precipitation. Mars inconjunct Jupiter and square the horizon was one indicator of dryness, and the other was Pluto square the meridian.

Heavy precipitation is, however, indicated in the lunar phase chart (middle wheel of Floods 6). Mars was conjunct the Midheaven and square Neptune, which was square the meridian, an aspect of heavy precipitation and strong thunderstorms. Also aspecting Neptune was Venus through a semisextile and parallel, and Venus was semisquare the Sun-Pluto conjunction (intense precipitation), in declination aspect to the lunar phase horizon, and sesquisquare/semisquare the ingress meridian. Jupiter square Uranus is also associated with thunderstorms, and this aspect was in effect with lunar phase Jupiter square ingress and lunar phase Uranus. Retrograde Mercury trine Saturn indicated precipitation and a strong weather front because lunar phase Mercury was square ingress Mercury.

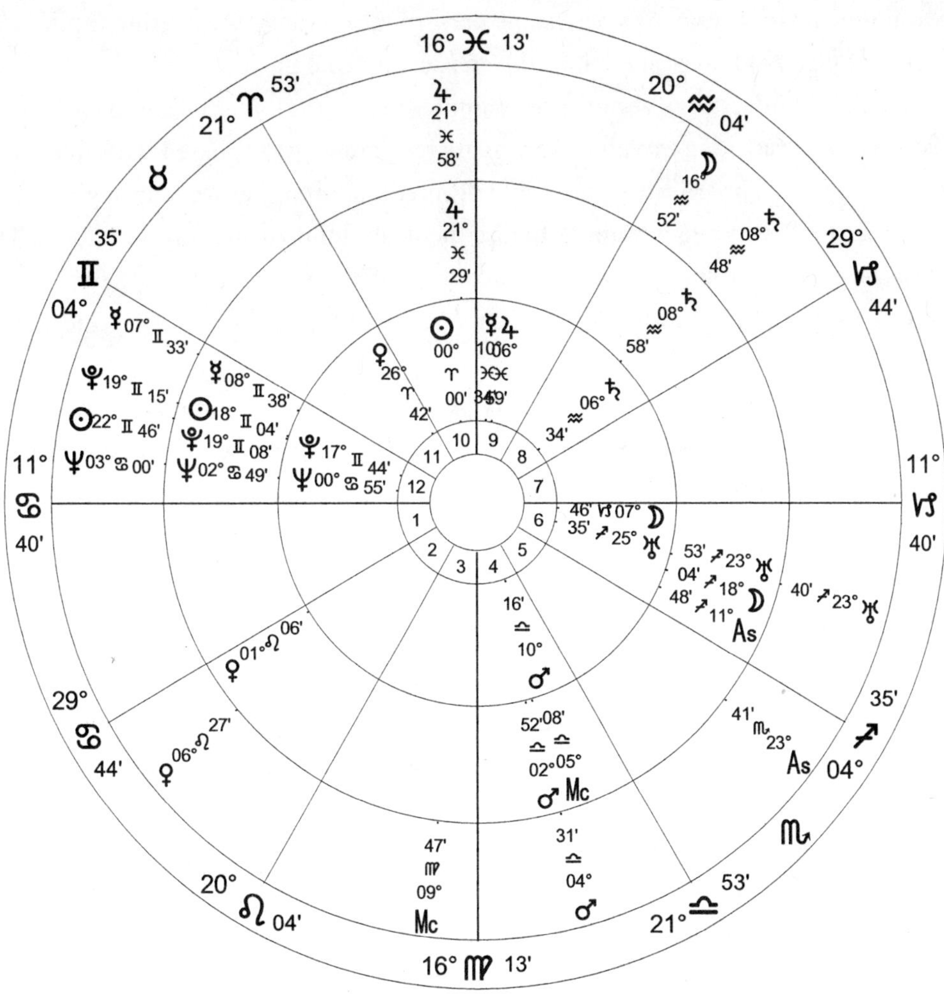

Floods 6

INNER WHEEL=*Spring Ingress / Natal Chart / March 21, 1903, Sat / 11:15 am PST +8:00*
Heppner, OR / 45°N21'12" 119°W33'24" / Geocentric / Tropical / Placidus / Mean Node

MIDDLE WHEEL=*Lunar Phase / Natal Chart / June 9, 1903, Tue / 7:07:54 pm PST +8:00*
Heppner, OR / 45°N21'12" 119°W33'24" / Geocentric / Tropical / Placidus / Mean Node

OUTER WHEEL=*Flood / Natal Chart / June 14, 1903, Sun / 5:15 pm PST +8:00*
Heppner, OR / 45°N21'12" 119°W33'24" / Geocentric / Tropical / Placidus / Mean Node

On the date of the flood (outer wheel of Floods 6), the Sun was opposition Uranus (storm, precipitation), Venus was sesquisquare lunar phase and transiting Jupiter and inconjunct ingress Jupiter (precipitation), and sextile/trine the lunar phase meridian. At the time the thunderstorm began, a few minutes before the water began to rush down the canyon, the transiting meridian was square retrograde lunar phase Mercury, activating those aspects. The transiting horizon triggered the Jupiter aspects through a trine/sextile that was exact a few minutes before the flash flood rushed through town. The third aspect was the transiting Moon, which formed a semisquare to lunar phase Mars in aspect with Neptune and the ingress Sun. The transiting Moon was also trine ingress Pluto.

Although it might have been difficult to predict a flash flood from these aspects—at least without making the connection between a dry spring and heavy precipitation—it would have been possible to predict a thunderstorm and its timing.

SALEM, OREGON, DECEMBER 1964

Salem was just one city affected by major flooding that impacted Oregon, coastal northern California, southwest Washington, Idaho, and Nevada. Seventeen people died in Oregon, and flooding from the Willamette River, which runs through Salem, covered almost 153,000 acres. The Willamette crested at 30 feet in the early morning hours of December 23, flooding hundreds of buildings; more than 1,000 people were rescued from flooded homes. Agricultural losses were also high, exceeding $10 million, sections of surrounding highways were under water, and more than thirty bridges were impassable. The Willamette crested again on December 24, at 29.5 feet.

The September 22, 1964, autumn ingress chart (inner wheel of Floods 7) had a key significator for abundant precipitation and flooding: Venus conjunct the Descendant and square Neptune, which was square the horizon. Any time you see this aspect aligned with the meridian or horizon, particularly when one (or both) planet is in a water sign, you can be sure there will be heavy precipitation during the season. Storms and heavy precipitation were also indicated by Venus semisextile Pluto, Jupiter square Saturn, and Mercury conjunction Uranus.

Temperatures were seasonal the week of the December 11, 1964, lunar phase (outer wheel of Floods 7), in the low 40s, until the temperature dropped to a high of 34 F on the 16th, followed by 19 F on the 17th; there was no rain on either of these days. The cold wave that began on the 16th is reflected in transiting retrograde Mercury sextile ingress and lunar phase Saturn, which were conjunct, and retrograde Mercury sesquisquare ingress Pluto and Mercury.

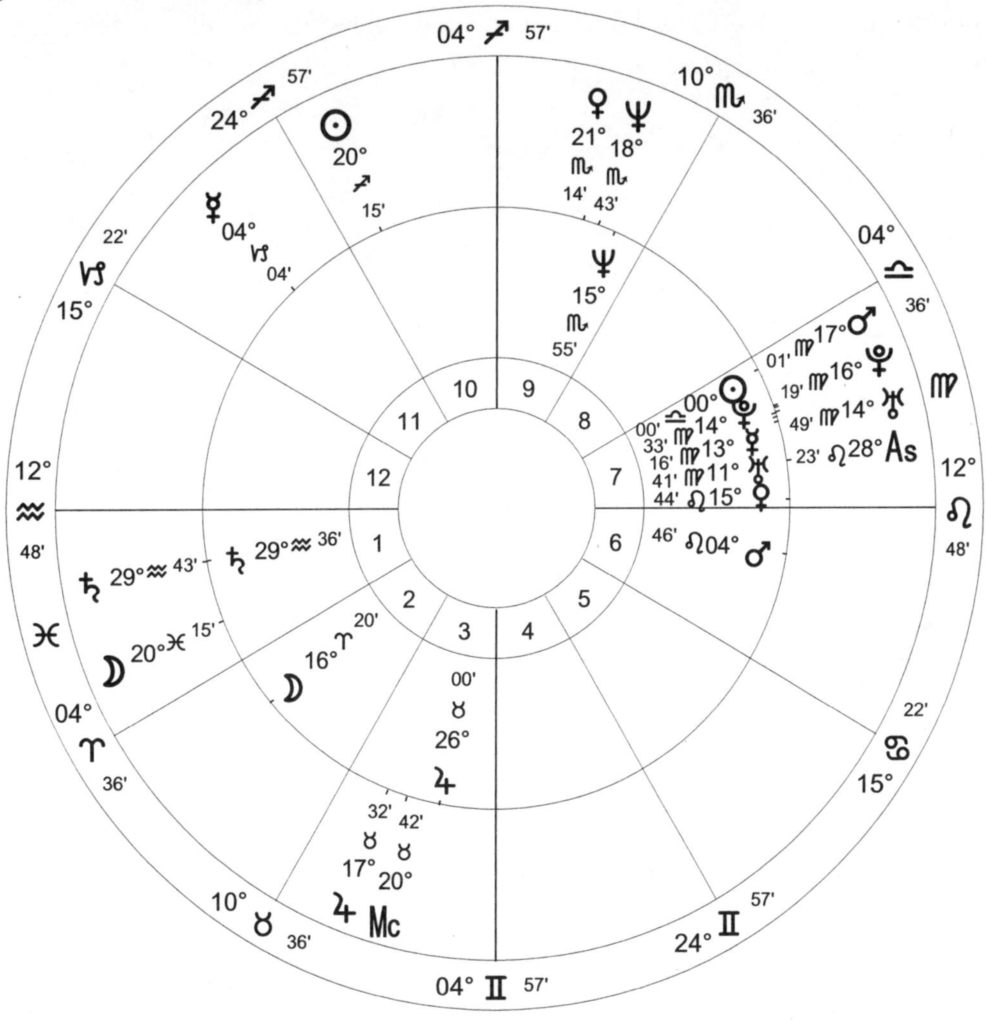

Floods 7

INNER WHEEL=*Autumn Ingress / Natal Chart / September 22, 1964, Tue / 5:16:38 pm PDT +7:00 Salem, OR / 44°N56'35" 123°W02'02" / Geocentric / Tropical / Placidus / Mean Node*

OUTER WHEEL=*Lunar Phase / Natal Chart / December 11, 1964, Fri / 10:01:21 pm PST +8:00 Salem, OR / 44°N56'35" 123°W02'02" / Geocentric / Tropical / Placidus / Mean Node*

In the December 11, 1964, lunar phase chart, Venus and Neptune were conjunct the IC and opposition Jupiter conjunct the Midheaven. But it wasn't until later in the week that this aspect was activated, resulting in heavy snow. The combination of frozen soil, which is unable to absorb much moisture, and heavy snow increased the flooding potential during the following week.

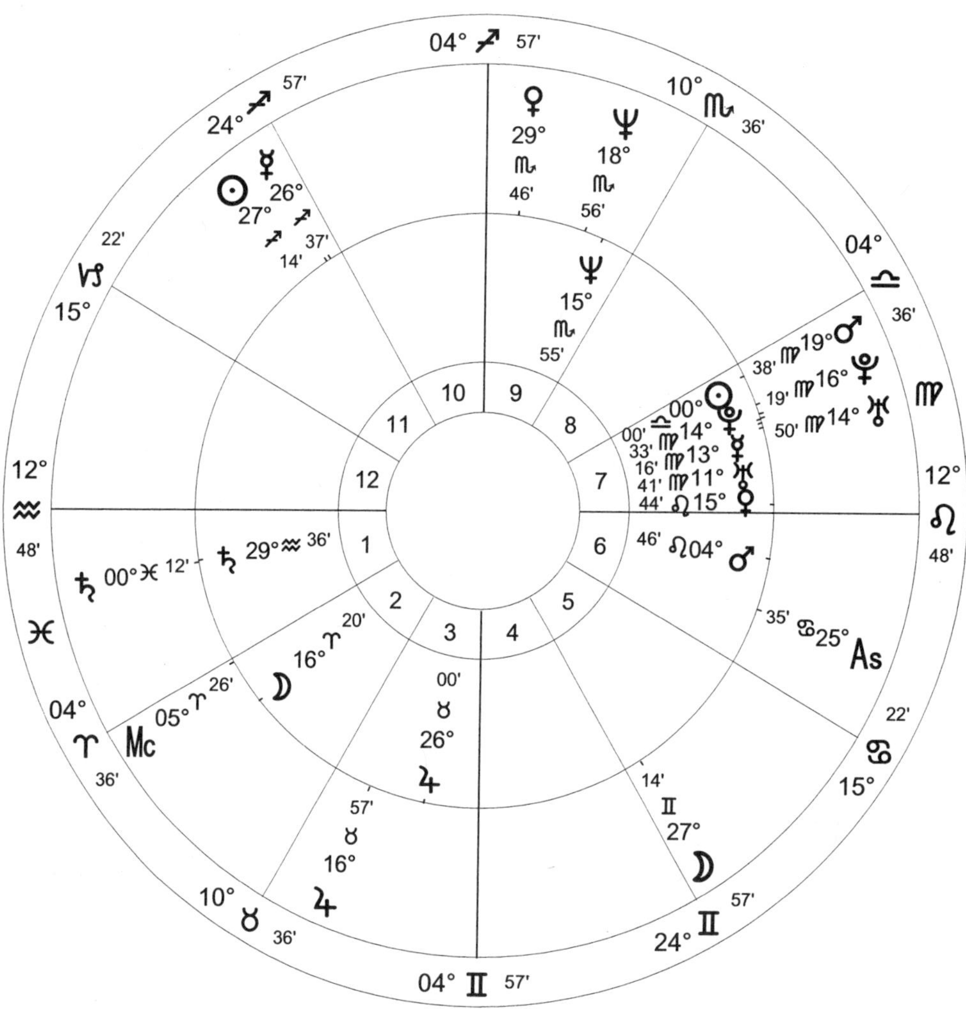

Floods 8

INNER WHEEL=*Autumn Ingress / Natal Chart / September 22, 1964, Tue / 5:16:38 pm PDT +7:00 Salem, OR / 44°N56'35" 123°W02'02" / Geocentric / Tropical / Placidus / Mean Node*

OUTER WHEEL=*Lunar Phase / Natal Chart / December 18, 1964, Fri / 6:41:11 pm PST +8:00 Salem, OR / 44°N56'35" 123°W02'02" / Geocentric / Tropical / Placidus / Mean Node*

As temperatures began to rise under the influence of the December 18 lunar phase (outer wheel of Floods 8), the snow began to melt and the rain began to fall. Temperatures and rainfall amounts on the days leading up to the first crest were December 18, 23 F, .09"; December 19, 37 F, 1.04"; December 20, 43 F, .74"; December 21, 55 F, 2.28"; and December 22, 61 F, 1.43".

At this lunar phase, the Sun was conjunct Mercury inconjunct/semisextile the horizon (precipitation), Mars was sextile Neptune (warmer), and Venus was square ingress and lunar phase Saturn (heavy precipitation). Transiting Venus advanced into Sagittarius, moving closer to the ingress meridian (conjunction to the Midheaven), and crossing it on the 21st and 22nd, the days with the highest levels of precipitation. At the same time, the transiting Sun moved into a sesquisquare with the ingress Venus-Descendant conjunction.

RAPID CITY, SOUTH DAKOTA, JUNE 9, 1972

During five hours on June 9, 1972, fifteen inches of rain fell on Rapid City, South Dakota. A strong flow of moist air from the Gulf continued to pump fuel into thunderstorms stalled west of the city. With the intense rain falling in an area of fewer than a hundred square miles, water surged down Rapid Creek, Canyon Lake quickly rose to twelve feet above its norm, and debris clogged the dam spillway. At 10:45 pm, the dam broke, sending a torrent of water rushing through the city, and by 12:15 am on June 10, the raging body of water in Rapid Creek was six and a half feet above flood stage.

The March 20, 1972, spring ingress chart (inner wheel of Floods 9) had Mars contraparallel Neptune (flash flood), an aspect that was aggravated by Jupiter in high declination semisextile Neptune. Venus parallel Saturn (heavy prolonged precipitation, overcast skies, low pressure, southerly air flow) was augmented by Saturn opposition and contraparallel Neptune (heavy precipitation, flash flood, low pressure, thunderstorms). All the factors necessary for intense rain were present, including the unusual weather indicated by Uranus semisquare Neptune.

Thunderstorms were also indicated by retrograde stationary Mercury opposition Uranus. Mercury was semisextile Venus (rain, southerly air flow) and semisquare Saturn (overcast skies, falling barometer, rain). Venus formed other rain aspects as well, with its inconjunct to Uranus and sesquisquare to Pluto. Mars and Neptune were contraparallel in high declination—the planetary configuration of a falling barometer and flash flooding.

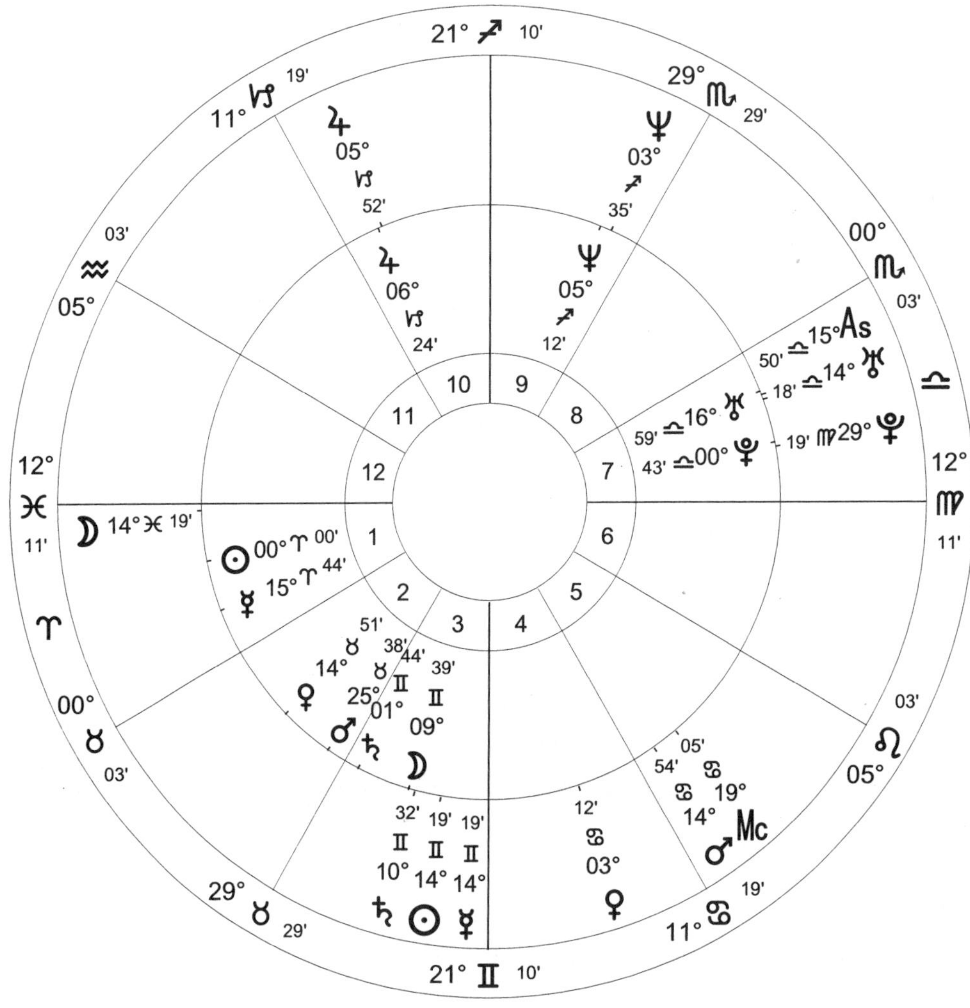

Floods 9

Inner Wheel=*Spring Ingress / Natal Chart / March 20, 1972, Mon / 5:22 am MST +6:00*
Rapid City, SD / 44°N04'50" 103°W13'50" / Geocentric / Tropical / Placidus / Mean Node

Outer Wheel=*Lunar Phase / Natal Chart / June 4, 1972, Sun / 3:22 pm MDT +5:00*
Rapid City, SD / 44°N04'50" 103°W13'50" / Geocentric / Tropical / Placidus / Mean Node

Neptune in the ninth house showed the precipitation coming from the west. Mercury was semisextile/inconjunct the horizon and Venus was sextile it. Jupiter, which was semisextile Neptune, formed a critical contact with the meridian with its contraparallel to the IC, and Uranus was parallel the Ascendant. The solar and lunar eclipses of January—25 Capricorn and 10 Leo—contacted the ingress angles.

By the time of the lunar phase of June 4, 1972 (outer wheel of Floods 9), Venus had turned retrograde and was in high declination, a harbinger of heavy precipitation. It formed a parallel with Mars (hot, humid, rain), an opposition with Jupiter in high declination (cloudy, excessive rain), an inconjunct with Neptune (hot, humid, cloudy, rain), and a square to stationary retrograde Pluto (hot, thunderstorms). Jupiter was parallel the IC.

Mars conjunct the Midheaven and square the horizon indicated thunderstorms with its contraparallel to Jupiter and square to Uranus, which was conjunct the Ascendant. Mars was also sesquisquare Neptune, which was semisquare/sesquisquare the horizon and the meridian (flash flood, falling barometer). The Sun conjunct and parallel Mercury (both in high declination) also indicated rain, while its contraparallel with Jupiter indicated thunderstorms. Intense rain accompanied by thunderstorms and a falling barometer were reflected in Sun conjunct Saturn, with the Sun contraparallel the IC and trine/sextile the horizon.

Lunar phase stationary retrograde Venus strongly aspected the ingress chart through an opposition with Jupiter (warm, cloudy, rain) and a semisquare to Venus (rain, cloudy). Jupiter was conjunct Jupiter, that planet having returned to its position in the ingress chart. The thunderstorms of Mars were activated by its semisquare to ingress Saturn and square to ingress Mercury and Uranus. Equally powerful was lunar phase Neptune forming contraparallels with ingress Venus (cloudburst, heavy rain, flood) and Saturn (above normal precipitation, low pressure, flash flood).

The horizon and meridian of both charts were similarly energized: ingress Mercury was conjunct the lunar phase Descendant, while aspects to the lunar phase Ascendant were formed by ingress Uranus (conjunction), Venus (inconjunct), and Saturn (sesquisquare). The lunar phase IC was aspected by ingress Mercury (square), ingress Venus (trine), ingress Saturn (sesquisquare), ingress Neptune (semisquare), and ingress Jupiter (parallel). Lunar phase Sun and Mercury were square the ingress horizon and parallel the ingress IC, and lunar phase Mars was trine/sextile the ingress horizon and parallel the ingress IC. Other lunar phase planets aspecting the ingress angles were Jupiter contraparallel the IC and the Sun parallel the IC.

Transiting aspects to the ingress chart on June 9, 1972, confirmed what was apparent at the ingress and lunar phase. Retrograde transiting Venus at 1 Cancer was square the ingress Sun (rain, hot, humid), semisquare ingress Venus (rain), parallel the IC, and parallel ingress Jupiter (warm, thunderstorms), which was also contraparallel the transiting Sun (warm, thunderstorms). Transiting Mars at 18 Cancer semisquare ingress Saturn and contraparallel ingress Jupiter indicated stronger thunderstorms, as transiting Neptune formed parallels with both ingress Venus and Saturn. Transiting Mars was parallel and Jupiter contraparallel the ingress IC.

In the five days between the lunar phase and when the rain began, transiting retrograde Venus moved into a contraparallel with lunar phase Jupiter (warm, cloudy, thunderstorms) and transiting Mercury a parallel with Venus. Transiting Mars formed a conjunction with the lunar phase Midheaven.

The transit-to-transit aspects of June 9, 1972, indicated more of the same, with the Sun at 19 Gemini semisextile Mars at 18 Cancer, and Mars sesquisquare Neptune at 3 Sagittarius. There were many parallels indicating rain and thunderstorms: Sun parallel Venus and Mars and contraparallel Jupiter; Mercury parallel Venus and Mars and contraparallel Jupiter; Venus contraparallel Jupiter and parallel Mars; and Mars contraparallel Jupiter.

DES MOINES, IOWA, 1993

Rain seemed as though it would never end in the Mississippi River Valley during the spring and summer of 1993, when many areas in Minnesota, Nebraska, Iowa, Kansas, Illinois, and Oklahoma received up to two feet of rain, much of it in July when it was too late for crops to be planted or replanted. Flooding extended to 10 million acres.

In Des Moines, Iowa, possibly the hardest hit major city, it was a case of "water, water everywhere, but not a drop to drink" when sewage treatment facilities were damaged by flooding in July. There, the Raccoon River topped out at more than nine feet above its official flood level of twenty-five feet. This major flood was a long-term event, reflected in two ingress charts.

The first indicator for heavy precipitation in the spring ingress chart (Floods 10) was retrograde Venus square Mars in Cancer, with both planets in northern declination (heavy rain). At best, the combination indicated heavy precipitation but not necessarily flooding. Venus was also contraparallel Saturn (heavy prolonged rain, especially with a Venus-Uranus aspect), square Uranus (heavy rain), and square Neptune (heavy rain, flash flood, cloudbursts). Retrograde planets always have a negative effect, even when in positive aspect.

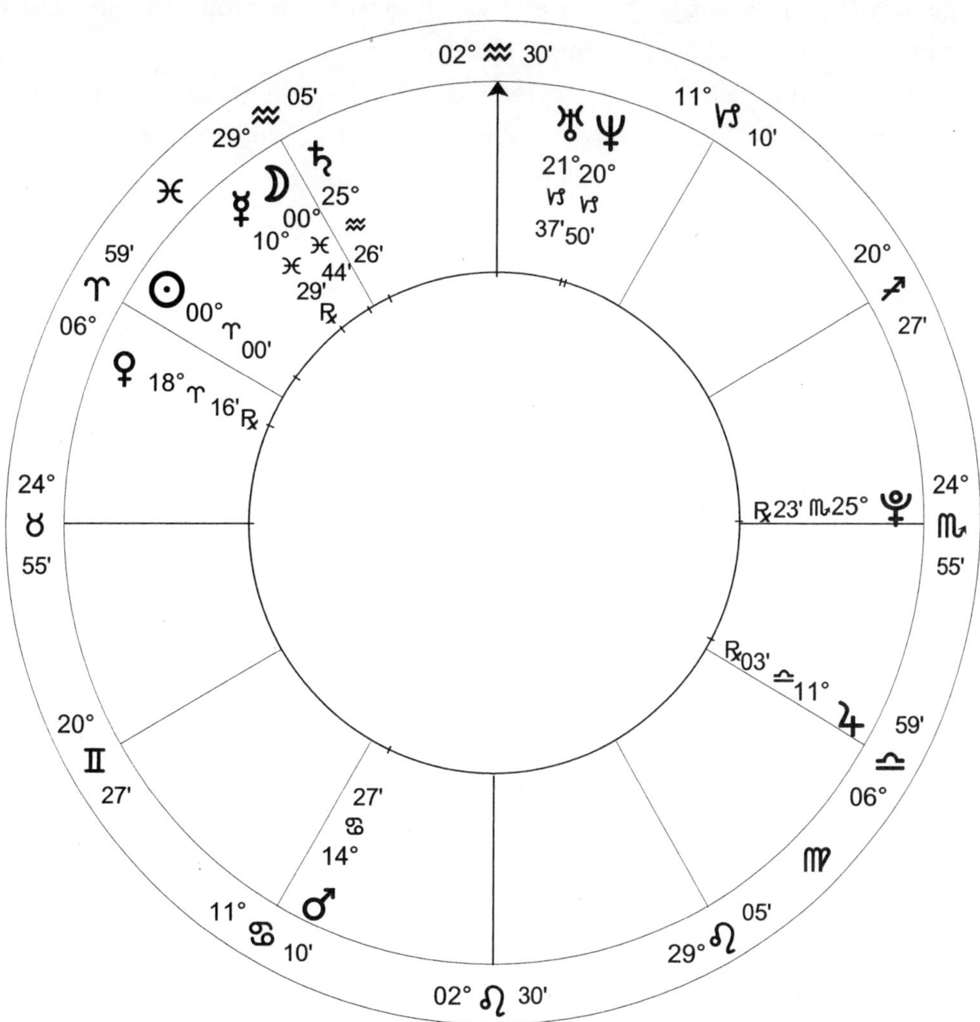

Floods 10: Spring Ingress / Natal Chart / March 20, 1993, Sat / 8:41 am CST +6:00
Des Moines, IA / 41°N36'02" 093°W36'32" / Geocentric / Tropical / Placidus / Mean Node

Another retrograde planet, Mercury, was trine Mars (thunderstorms and westerly winds), inconjunct retrograde Jupiter (northerly air, precipitation), and semisquare Uranus (thunderstorms).

Mars square Jupiter and Jupiter semisquare Pluto (thunderstorms) combined their force with Jupiter sesquisquare Saturn (above-normal precipitation), Saturn square Pluto (windy storm), and Saturn semisextile Neptune (rain).

Jupiter was sesquisquare/semisquare the horizon, and Saturn was square the horizon. The lunar aspects to Mercury (parallel), Venus (semisquare), and Mars (sesquisquare) helped activate those planets through the Moon's inconjunct/semisextile to the meridian. The lunar eclipse of December 9, 1992, at 18 Gemini was semisquare/sesquisquare the meridian, and the solar eclipse of December 24, 1992, at 2 Capricorn was inconjunct/semisextile the meridian.

As is often true with unusual weather, Uranus and Neptune were in aspect—conjunct and parallel—and in the ninth house, an alert that weather would come from the west. They were trine/sextile the horizon, activating that latitude, but it was the Ascendant-IC parallel that brought the Mississippi Valley into high focus. Without that one important aspect, it would have been easy to predict severe rain to the west instead of centered in the Midwest.

As April unfolded, the spring ingress chart was first activated when transiting Mars in Cancer opposed Uranus and Neptune and formed a square with Venus. Saturn, transiting the late degrees of Aquarius, was moving closer to a sesquisquare with Mars, which would become exact in early May. At the end of April, transiting Venus turned direct at 3 Aries conjunct the ingress Sun and opposition transiting Jupiter at 7 Libra, and transiting Mercury reached 0 degrees declination on April 20, unleashing its negative side. The transiting Sun formed a contraparallel with Saturn on April 24, and Mars's contraparallel to Uranus was exact two days later.

May began with transiting Mars in the early degrees of Leo crossing the ingress IC, while transiting Venus formed a nearly month-long contraparallel to ingress Pluto, which was conjunct the ingress Descendant. Jupiter was at 0 degrees declination all month. The lunar phase of May 21, 1993 (solar eclipse at 0 Gemini), was square transiting Saturn at 0 Pisces, which formed a conjunction to the ingress Moon. Near the end of May, transiting Venus at 20 Aries formed a square with ingress Neptune.

In June, Jupiter continued its stay at 0 degrees for the first three weeks as Mars continued through Leo, opposing ingress Saturn and forming a square with ingress Pluto and the horizon. This activated transiting stationary retrograde Saturn's conjunction with the ingress Moon. The transiting Sun moved on to contraparallel ingress Uranus all month and for three weeks was parallel transiting Mercury, which was parallel ingress Mars early in the month and opposition ingress Uranus and Neptune mid-month.

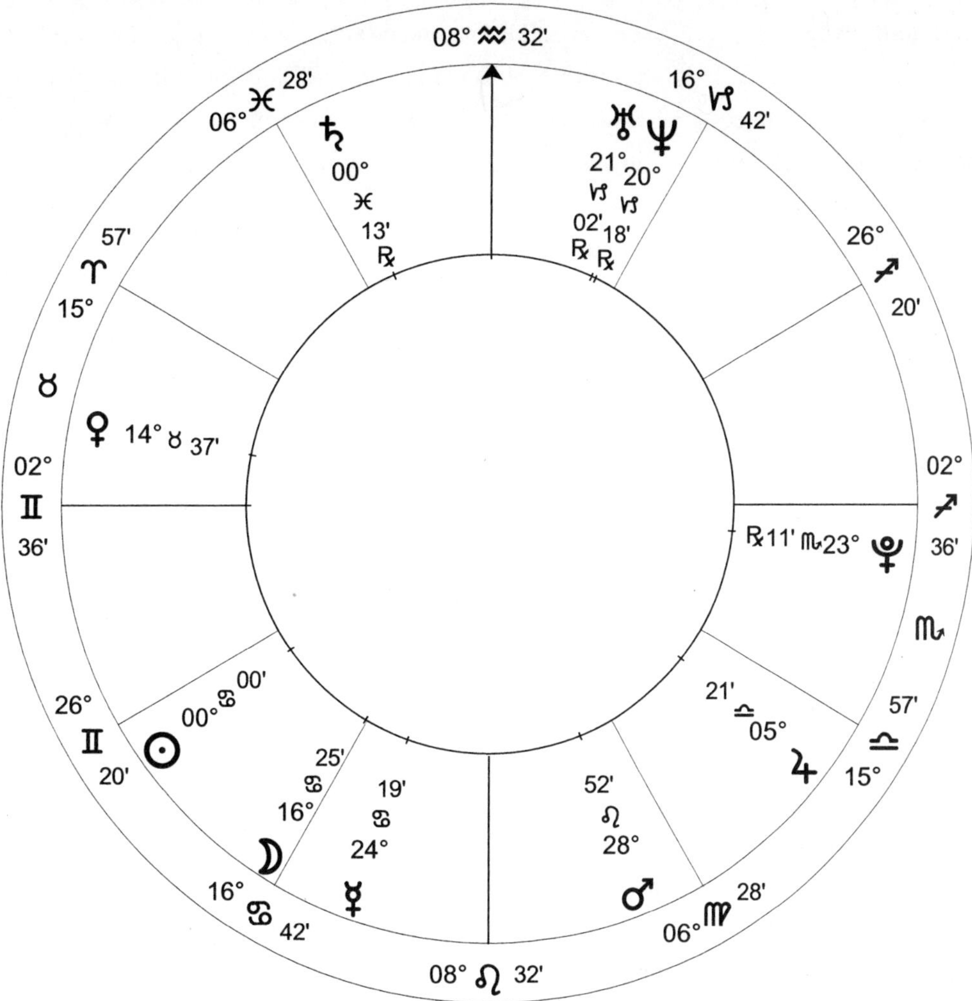

Floods 11: Summer Ingress / Natal Chart / June 21, 1993, Mon / 4:00 am CDT +5:00
Des Moines, IA / 41°N36'02" 093°W36'32" / Geocentric / Tropical / Placidus / Mean Node

By the summer 1993 ingress (Floods 11), Midwesterners were beginning to wonder if they would ever see the end of rain and a glimmer of typical sunny summer weather. The ingress chart could make only one promise: more rain. The solar eclipse of May 21, 1993, at 0 Gemini was conjunct the Ascendant.

Uranus and Neptune were still conjunct in the ninth house, bringing rain from the west, and sesquisquare and contraparallel the Ascendant. Mercury, parallel the Ascendant, was contraparallel and opposition both planets, an indicator of thunderstorms.

The Sun was semisquare Venus (rain), sextile Mars (heat), semisextile/inconjunct the horizon, and trine Saturn (rain). The Venus-Saturn connection with the Sun was further energized by Venus contraparallel Saturn and parallel Mars (heavy rain). Saturn was again activated by its opposing planet, Mars, which also formed a contraparallel (windy storm). Mercury trine Pluto furnished more westerly winds to bring storms to the Midwest. The Moon, semisquare/sesquisquare the horizon, aspected these key planets: Venus (sextile), Mars (semisquare), Saturn (sesquisquare), and Neptune (opposition).

Saturn opposition Mars, with both planets square the horizon, indicated continuing wet weather in the Mississippi River Valley. Saturn in Pisces was associated with the prolonged rain, and Mars in Leo added heat to generate thunderstorms.

The Mars-Saturn aspects were exact a few days after the ingress as Mars moved forward into Virgo. The separation of that aspect might have ended the rain, except that Saturn, in retrograde motion, returned to Aquarius, bringing the ingress aspect closer and closer to exact. When it reached that point on July 24, five inches of rain fell on central Iowa.

Mercury aspects throughout the month of July contributed to the ongoing downpour. As that planet turned retrograde on July 2 at 28 Cancer, it formed a semisextile with ingress Mars and an inconjunct with ingress Saturn the same day. Turning direct on July 26 at 18 Cancer, it was conjunct the ingress Moon and opposition ingress Uranus and Neptune at the time of the five-inch rain at the end of the month. In mid-July, retrograde transiting Mercury in Cancer opposed ingress Uranus and Neptune at the same time that the transiting Sun was contraparallel those planets and parallel ingress Mercury.

Transiting Venus in Gemini formed a conjunction/opposition with the ingress horizon and a square to ingress Saturn in early July. As it continued forward, Venus was contraparallel Uranus and Neptune the last half of the month, and parallel ingress Mercury the third week. Throughout the month at various times, the transiting Sun was parallel or contraparallel ingress and transiting Mercury, ingress and transiting Uranus and Neptune, transiting Venus, and the ingress Moon.

Uranus and Neptune gradually narrowed their orb of conjunction all month, an aspect that would become exact in mid-August. But by then, the parallel aspect had separated, and the planets lost some of their power. At the same time, the all-important summer ingress planet, Saturn, was retrograding away from its exact opposition to ingress Mars.

CHAPTER 7

Temperature

The sun is at its highest point in the summer in the Northern Hemisphere, so its rays are more concentrated and focused, more direct, than they are in the winter, when light strikes Earth at a shallow angle. The sun remains visible for longer periods of time in the summer (the longest day of the year is the day of the summer ingress, and the shortest is the winter ingress), thus giving it more time to heat the earth each day. As summer turns to fall and winter, the sun shines in the Northern Hemisphere for fewer and fewer hours, causing the earth to lose its retained heat faster than it gains new heat. The reverse happens when winter turns to spring and summer; days gradually lengthen and the sun shines for more hours.

In addition to these seasonal factors, high-pressure systems often contribute to extended periods of high heat in the summer months. These air masses can remain "parked" over a vast area (many states) for several days or even weeks, continuing to build heat energy.

The planets of heat—Venus, Mars, Jupiter, Neptune, and Pluto—are always prominent in ingress and lunar phase charts that indicate high temperatures. When they are in combination with each other, excessively high heat generally ensues. Uranus can also be associated with high heat in summer because it is a planet of high pressure, as is Jupiter. Other seasons are sometimes unseasonably warm when these planets dominate at the ingress.

The cold weather planets—Saturn, Uranus, and often Mercury—are invariably present in ingress and lunar phase charts that reflect lower temperatures. As in other intense weather systems, Pluto can function to intensify the cold rather than bring warmer temperatures. When Saturn and Uranus are combined, temperatures are often unseasonably cold, and this can be amplified when Jupiter is also involved. An emphasis of cold planets in generally warmer seasons indicates unseasonably cool temperatures.

Aquarius nearly always indicates very cold temperatures, and the same is often true with Capricorn. Mercury retrograde in Aquarius is associated with windy, cold weather, and Venus in Aquarius can reflect icy temperatures.

Another important consideration when forecasting heat and cold is declination. When a significant number of planets—six or seven or more—are in southern declination, cold temperatures can be expected in the Northern Hemisphere; the reverse is true when many planets are in northern declination—temperatures are generally warmer.

The Sun, Mercury, and Venus are always so close together that they are all in southern declination during the winter months and in northern declination during the summer. The change in declination generally happens within a few weeks of when the Sun enters north or south declination (at the ingress), depending on the retrograde patterns of Mercury and Venus. As a result, the impact of the slow-moving outer planets in long-term weather trends is of even greater significance.

When many planets are in southern declination during winter months, the trend is for colder weather; this is especially true when the planets are also in longitudinal aspect (conjunction, opposition, etc).

The Moon, Mars, and Jupiter, which transit more rapidly, can be key factors in flipping temperatures one way or the other, depending on whether they tip the balance to southern or northern declination. Signs of northern declination are Aries, Taurus, Gemini, Cancer, Leo, and Virgo; southern declination signs are Libra, Scorpio, Sagittarius, Capricorn, Aquarius, and Pisces.

Pinpointing a specific temperature is unrealistic with astrometeorology. Rather, think in terms of temperature range: seasonal, unseasonal, above average, below average, etc. If you see few or no planetary influences that would indicate temperatures below or above average, conditions will be seasonal. Again, it's important to learn the climatology of the location. You can find daily averages for most locations at www.wunderground.com.

Phoenix, Arizona, 1990 and 2013

Although the desert Southwest is accustomed to extreme temperatures on the hottest summer days, there was a new high in Phoenix on June 26, 1990: 122 F. Air travelers found themselves stranded when the airport was forced to close, because aviation tables listing parameters necessary for takeoff stopped at 120 F. Motorists experienced overheated engines, and residents and power companies hoped the high demand for energy and stress on air conditioners would survive the strain.

The summer ingress chart (inner wheel of Temperature 1) promised unseasonably high temperatures, with Mars in Aries square Jupiter and Neptune and inconjunct Pluto. Mars and Jupiter were in northern declination, further intensifying the planets' natural tendencies. Pluto was heavily aspected: Sun sesquisquare, Mars inconjunct, Jupiter trine, and Neptune sextile. Jupiter was in opposition and contraparallel to Neptune, and Venus was parallel and the Sun sextile the Midheaven. Both the meridian and the horizon were aspected by the solar eclipse of January 26, 1990, at 7 Aquarius.

The planetary pattern was nearly the same the next day when the lunar phase occurred (outer wheel of Temperature 1), aggravating the ingress indicators. When an ingress and a lunar phase occur very close together, the indicated weather pattern is usually intensified. The Mars, Jupiter, and Neptune aspects were still in effect, and the Sun and Jupiter were contraparallel the IC. Aspecting the meridian by trine/sextile was the lunar eclipse of February 9, 1990, at 21 Leo.

On June 26, 1990, transiting Venus at 2 Gemini was semisquare ingress Jupiter and Mars. The transiting Moon in late degrees of Leo was contraparallel ingress Mars, square ingress Venus, and semisquare ingress Jupiter. The same transiting aspects occurred with the lunar phase chart.

This event demonstrates the power and importance of parallels in astrometeorology. Pluto in the fourth house of the ingress chart promised a hot summer to come in the West, but it was only through a parallel that the lunar phase chart indicated which week would be one of excessive heat; Pluto was contraparallel the Ascendant. To further indicate that Phoenix would experience intense heat, Jupiter and the lunar phase Sun were contraparallel the lunar phase IC, and Venus was parallel the ingress Ascendant. Also significant was how the two charts were linked: the lunar phase meridian and horizon aspect the ingress meridian and horizon by semisquare/sesquisquare.

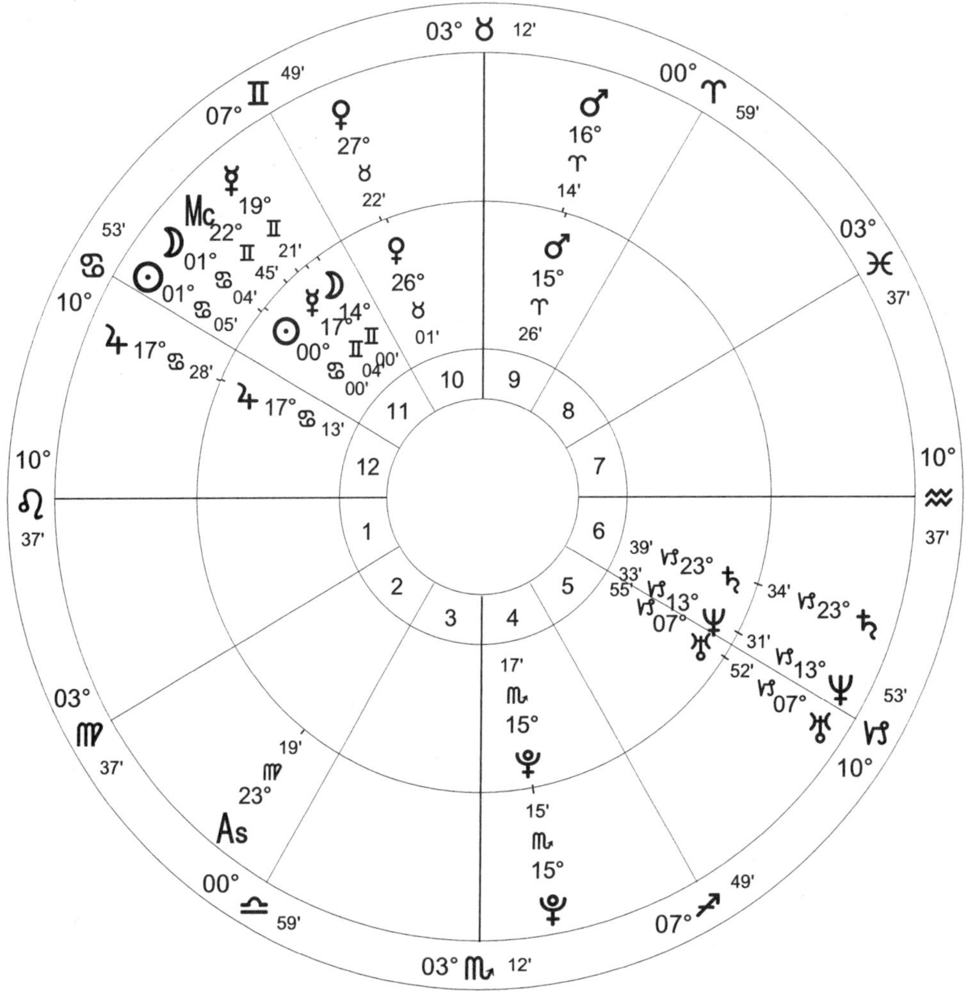

Temperature 1
INNER WHEEL=*Summer Ingress / Natal Chart / June 21, 1990, Thu / 8:34 am MST +7:00
Phoenix, AZ / 33°N26'54" 112°W04'24" / Geocentric / Tropical / Placidus / Mean Node*

OUTER WHEEL=*Lunar Phase / Natal Chart / June 22, 1990, Fri / 11:54 am MST +7:00
Phoenix, AZ / 33°N26'54" 112°W04'24" / Geocentric / Tropical / Placidus / Mean Node*

Further indicators of high heat in Phoenix were ingress aspects to the angles: Sun trine/sextile the meridian, and Uranus (high pressure) and Pluto inconjunct/semisextile the horizon. In the lunar phase chart, the Sun was contraparallel and Uranus parallel the IC.

Twenty-three years later, Phoenix again experienced an extremely high temperature of 118 F. The Sun, Mars, Jupiter, and Pluto were involved in the summer ingress charts of both 1990 and 2013, but Uranus had a much stronger presence in the 2013 event (inner wheel of Temperature 2). In 2013, Uranus was sextile Mars (heat), square Pluto (high pressure, excessive heat or cold), and sextile/trine the horizon; Pluto was semisquare and parallel the Midheaven and semisextile the Ascendant. Other heat aspects at the ingress were Venus sesquisquare Neptune, with Venus trine/sextile the meridian and Neptune semisextile/inconjunct the horizon; and Venus, Mars, and Jupiter all in parallel aspect. There were seven planets in northern declination.

Lunar phase aspects (middle wheel of Temperature 2) for June 23 were mostly the same because only a few days separated the two dates. However, Neptune was conjunct the Midheaven, trine the Sun, and sesquisquare Mercury (these aspects also indicate the start of the monsoon season, which brings a southerly flow of moist, humid air). Lunar phase Uranus was semisquare/sesquisquare the meridian (high pressure), Jupiter was trine/sextile the meridian (high pressure), and the Sun was conjunct and parallel Jupiter (heat).

The temperature climbed all week and culminated on June 29 with 118 F. On that date, transiting Mars was conjunct the June 23 lunar phase Ascendant, Venus was semisextile Jupiter and sesquisquare/semisquare the lunar phase horizon, and the Sun was opposition Pluto (high heat). The parallels in effect at the ingress and lunar phase were still active in the transits.

From a meteorology perspective, this high-temperature event was the result of a closed high-pressure system over the western states. A closed high is an uncommon event and is distinguished from a "normal" high by its shape. On meteorology maps, high- and low-pressure systems are seen as large, loopy waves. When the wave looks like the top half of a circle or an oval, high pressure is in effect; when the wave looks like the bottom half of a circle or an oval, low pressure is in effect. These systems usually move in a wave-like pattern across the continent. When a closed high occurs, however, it is seen as a closed circle or oval on a meteorology map, and the high pressure is focused or concentrated over an area for an extended time (several days to a week or more).

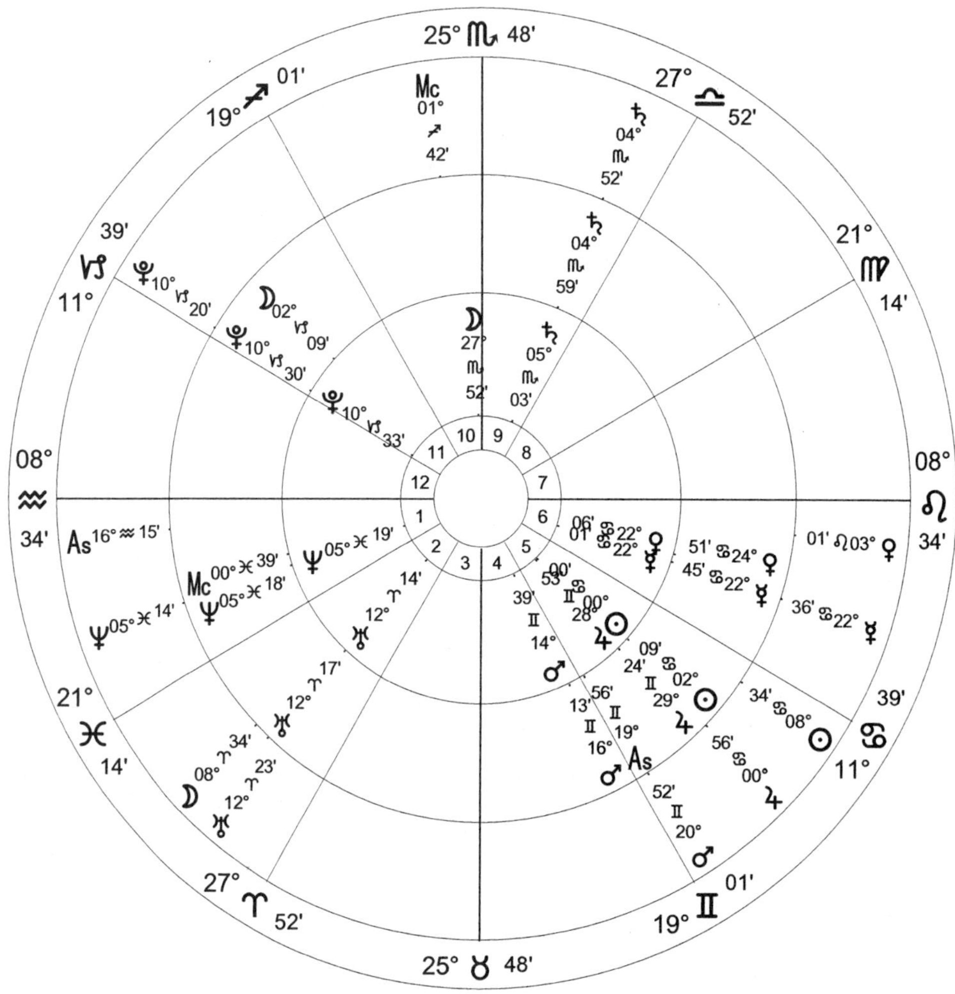

Temperature 2

INNER WHEEL=*Summer Ingress / Natal Chart / June 20, 2013, Thu / 10:04 pm MST +7:00
Phoenix, AZ / 33°N26'54" 112°W04'24" / Geocentric / Tropical / Placidus / Mean Node*

MIDDLE WHEEL=*Lunar Phase / Natal Chart / June 23, 2013, Sun / 4:32 am MST +7:00
Phoenix, AZ / 33°N26'54" 112°W04'24"/ Geocentric / Tropical / Placidus / Mean Node*

OUTER WHEEL=*Lunar Phase / Natal Chart / June 29, 2013, Sat / 9:53 pm MST +7:00
Phoenix, AZ / 33°N26'54" 112°W04'24"/ Geocentric / Tropical / Placidus / Mean Node*

The closed high for the 2013 event is reflected in the lunar phase chart through the Sun-Jupiter conjunction trine/sextile the meridian, and the Mars-Uranus sextile, with Mars conjunct the horizon. In the 1990 summer ingress chart, the Sun-Jupiter parallel was contraparallel Uranus, with Uranus trine/sextile the meridian. (A closed low can also occur, and this usually marks an area of extreme low pressure associated with a storm.)

SUMMER 2011 HEAT WAVE

Areas from the Plains states to the East Coast experienced the hottest summer on record in 2011. The heat wave began in mid-July and extended into the second week of August. Many records were set, including in Minnesota, with a record high minimum of 80 F; Arkansas, with twenty consecutive days of 100 F; and Texas, which had the greatest number of 100 F days in a calendar year. Kansas, where the average summer temperature is 101.2 F, saw 113 F on July 27 in Hutchinson and Salina, and Hartford, Connecticut, set a new all-time record high of 103 F on July 22.

Much of the heat wave was reflected in a Uranus-Pluto square (high pressure, heat), but Venus, Mars, and Neptune were also prominent. At the ingress, there were thirteen aspects indicative of excessive heat in addition to six planets in northern declination. The aspects were Mars semisextile Jupiter, sextile Uranus, and square Neptune; Venus semisquare Jupiter; Uranus square Pluto and semisextile Neptune; Sun semisextile Mars, sextile Jupiter, and trine Neptune; and Jupiter semisextile Uranus, sextile and contraparallel Neptune, and trine Pluto. The number of aspects explains why such a massive area was affected, because several of the aspects made connections with many longitudes and latitudes across the continent. We'll look at two examples, one in the Midwest and the other on the East Coast.

The June 21, 2011, summer ingress chart for Salina, Kansas (inner wheel of Temperature 3), showed potential for high temperatures throughout much of the summer. Venus aspected both the meridian and horizon through a conjunction and square, and Jupiter was semisquare the Midheaven and sextile the Sun. Note Mars at 0 Gemini, west of Salina at the ingress. Mars was conjunct the ingress Midheaven in Colorado, another area that experienced high temperatures during the summer of 2011. Using Mars, you could have tracked the heat progression as this planet transited through Gemini and then Cancer, contacting longitudes all the way to the east during the succeeding two months.

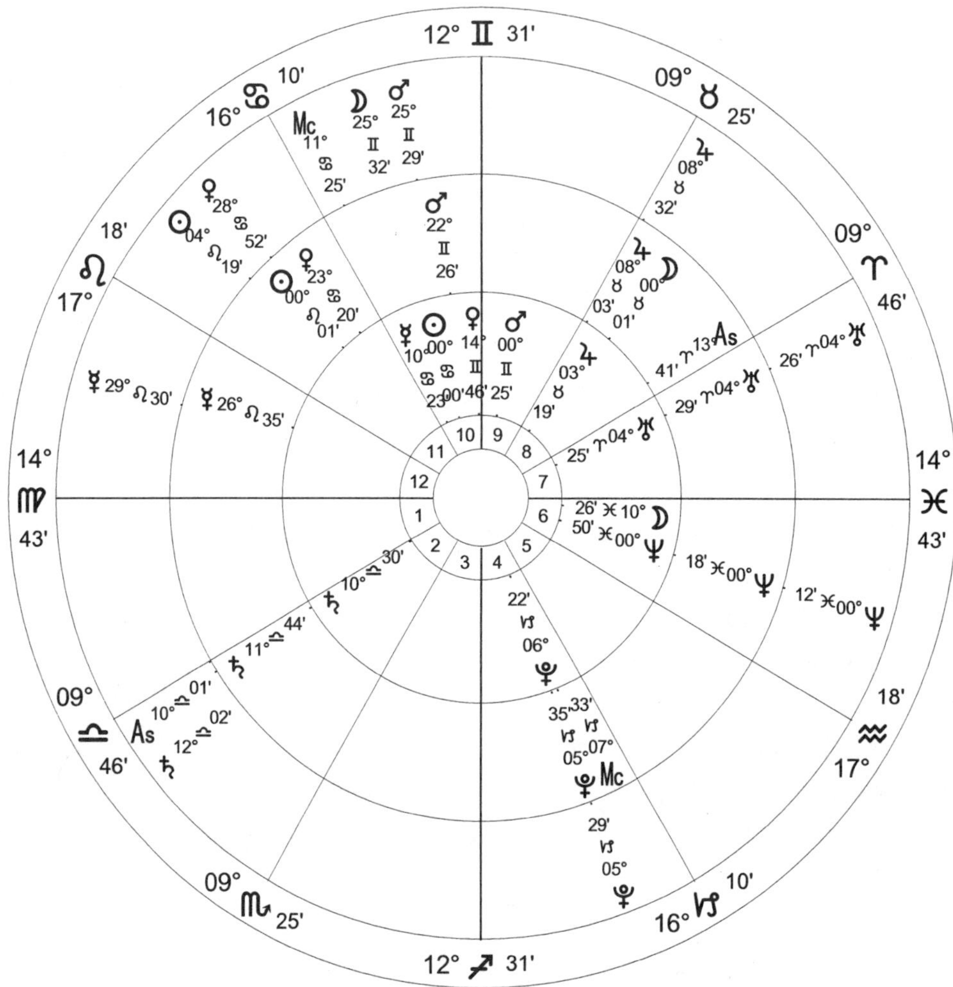

Temperature 3

Inner Wheel=*Summer Ingress / Natal Chart / June 21, 2011, Tue / 12:16:30 pm CDT +5:00*
Salina, KS / 38°N50'25" 097°W36'40" / Geocentric / Tropical / Placidus / Mean Node

Middle Wheel=*Lunar Phase / Natal Chart / July 23, 2011, Sat / 0:01 am CDT +5:00*
Salina, KS / 38°N50'25" 097°W36'40" / Geocentric / Tropical / Placidus / Mean Node

Outer Wheel=*High Heat / Natal Chart / July 27, 2011, Wed / 12:00 pm CDT +5:00*
Salina, KS / 38°N50'25" 097°W36'40" / Geocentric / Tropical / Placidus / Mean Node

The temperature in Salina was 113 F on July 27, under the lunar phase chart (middle wheel of Temperature 3), which had Uranus and Pluto conjunct and square the meridian. The angle was also aspected by a trine/sextile from Jupiter, which was semisquare Mars. Ingress Venus was sextile/trine the lunar phase horizon. The solar eclipse of June 1, 2011, at 11 Gemini was conjunct the ingress meridian and square the ingress horizon, and the solar eclipse of July 1, 2011, at 9 Cancer was conjunct/opposition the lunar phase horizon.

On the date of the season's highest temperature, July 27 (outer wheel of Temperature 3), the transiting Moon formed a conjunction with lunar phase and transiting Mars as the thermometer headed for 113 F. Transiting Mercury was sextile the ingress Sun, activating the solar aspects, as well as the lunar phase horizon through a semisquare/sesquisquare. Also reflective of high heat was transiting Mars parallel ingress Venus and transiting Venus parallel ingress Mars.

Several days before the Salina high heat, the thermometer soared in Hartford, Connecticut, when the temperature reached a record high of 103 F on July 22 (outer wheel of Temperature 4). The July 1, 2011, solar eclipse at 9 Cancer aspected the ingress meridian and horizon through a conjunction and square (inner wheel of Temperature 4).

Ingress aspects were the same in the Hartford chart, but there, the Uranus-Pluto square was aligned with both the meridian and horizon.

Mars in the Hartford July 15 lunar phase chart (middle wheel of Temperature 4) was trine/sextile the meridian, parallel the horizon, and sextile Mercury, which was sesquisquare the Uranus-Pluto square (both lunar phase and ingress) and conjunction-opposition the lunar phase meridian. Lunar phase Venus was sesquisquare Neptune and inconjunct/semisextile the meridian and horizon. Jupiter was semisextile Uranus, contraparallel Neptune, and trine Pluto, all indicators of high heat. The meridians in the ingress and lunar phase charts were sesquisquare.

(Mercury conjunct the Midheaven or IC often indicates cooler, windy weather, but this obviously wasn't true later in the week, and in any case it would be less so with Mercury in Leo, a fire sign. However, the high temperature on the day of the ingress was a seasonably pleasant 84 F. The next day, as Mercury moved away from the IC, the high was 90 F. It wasn't until the 21st that the temperature rose into the high 90s.)

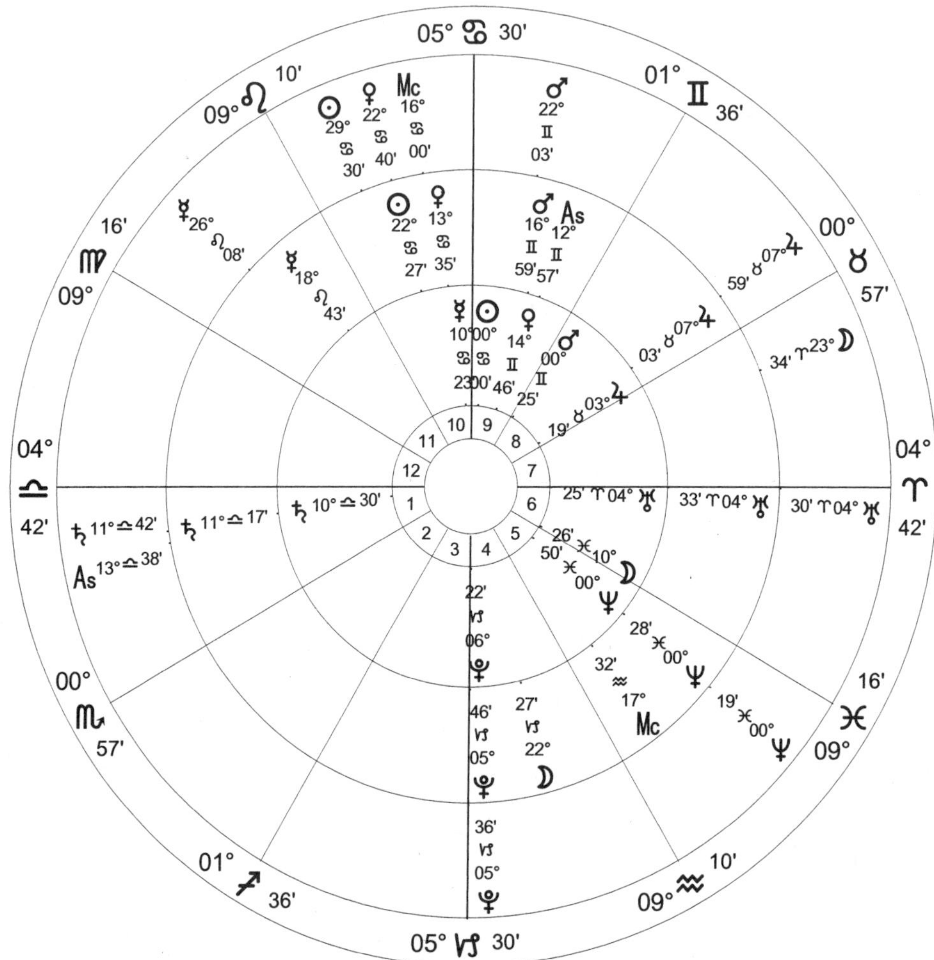

Temperature 4

Inner Wheel=*Summer Ingress / Natal Chart / June 21, 2011, Tue / 1:16:30 pm EDT +4:00*
Hartford, CT / 41°N45'49" 072°W41'08" / Geocentric / Tropical / Placidus / Mean Node

Middle Wheel=*Lunar Phase / Natal Chart / July 15, 2011, Fri / 2:39:36 am EDT +4:00*
Hartford, CT / 41°N45'49" 072°W41'08" / Geocentric / Tropical / Placidus / Mean Node

Outer Wheel=*High Heat / Natal Chart / July 22, 2011, Fri / 12:00 pm EDT +4:00*
Hartford, CT / 41°N45'49" 072°W41'08" / Geocentric / Tropical / Placidus / Mean Node

On July 22, the transiting Sun was inconjunct ingress Neptune, transiting Mercury was sesquisquare the ingress Ascendant and Midheaven, and transiting Mars was semisquare ingress Jupiter. Transiting aspects to the lunar phase chart also indicated high heat: Mars parallel the horizon and semisextile the Sun, which was parallel the horizon; Venus conjunct Sun; and Venus parallel Mars and the horizon.

RIO GRANDE, TEXAS, MARCH 31, 1954

The high heat of March 31, 1954, was unexpected and unusual in Rio Grande, Texas, even though it's one of the southernmost cities in the United States. The temperature reached 108 F a little more than a week after the spring ingress of March 21.

The ingress chart (inner wheel of Temperature 5) had strong aspects for high temperatures. Mars opposed Jupiter, an aspect that was intensified because the two were contraparallel. Mars was also trine Pluto, which was parallel Jupiter and sextile Neptune. Jupiter was parallel Uranus and sesquisquare Mercury (high pressure), while Venus, Jupiter, Uranus, and Pluto in northern declination also indicated a higher range of temperature for the season.

Pluto focused its energy on the ingress longitude, as did Venus and Jupiter sextile/trine the meridian. Venus also was inconjunct/semisextile the horizon. The January 5, 1954, solar eclipse at 14 Capricorn was sextile/trine the horizon and semisextile/inconjunct the meridian.

The lunar phase chart of March 27, 1954 (inner wheel of Temperature 5), retained the Mars-Jupiter and Mars-Pluto configurations and the Jupiter-Pluto sextile, and added Mars sextile Neptune and Jupiter sextile Pluto. One significant factor that indicated an elevated temperature was numerous Venus aspects: Venus square Uranus, contraparallel Neptune, trine Mars, sextile Jupiter, and trine Pluto. Both January 1954 eclipses aspected the lunar phase meridian: the January 5 solar eclipse at 14 Capricorn (sesquisquare/semisquare) and the January 19 lunar eclipse at 28 Cancer (semisextile/inconjunct).

Aspects between the ingress and lunar phase charts were also indicative of heat. The lunar phase Sun was sesquisquare ingress Pluto, lunar phase Venus was trine ingress Pluto, and ingress Venus was sesquisquare/semisquare the lunar phase meridian and sextile/trine the lunar phase horizon.

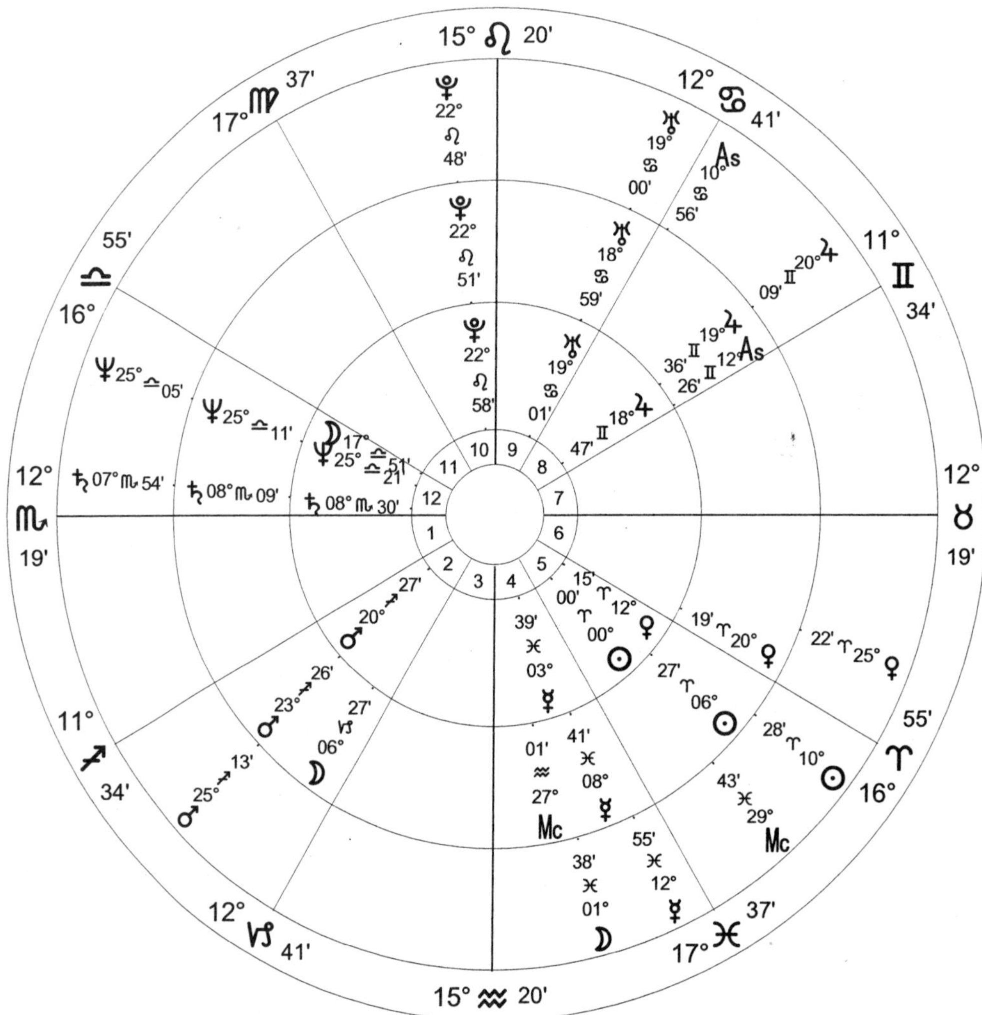

Temperature 5
INNER WHEEL=*Spring Ingress / Natal Chart / March 20, 1954, Sat / 9:54 pm CST +6:00*
Rio Grande City, TX / 26°N22'46" 098°W49'12" / Geocentric / Tropical / Placidus / Mean Node

MIDDLE WHEEL=*Lunar Phase / Natal Chart / March 27, 1954, Sat / 10:14 am CST +6:00*
Rio Grande City, TX / 26°N22'46" 098°W49'12" / Geocentric / Tropical / Placidus / Mean Node

OUTER WHEEL=*High Heat / Natal Chart / March 31, 1954, Wed / 12:00 pm CST +6:00*
Rio Grande City, TX / 26°N22'46" 098°W49'12" / Geocentric / Tropical / Placidus / Mean Node

Transiting aspects on March 31 (outer wheel of Temperature 5) activated the ingress chart. The transiting Sun at 11 Aries was conjunct ingress Venus, transiting Venus at 25 Aries was opposition and parallel ingress Neptune, and transiting Jupiter at 20 Gemini had moved to within 19 minutes of exactitude in its opposition to ingress Mars. The transiting Sun was parallel ingress Venus, and transiting Mars was contraparallel ingress Pluto. The transiting Moon in late Aquarius and early Pisces set off the ingress Moon, Venus, Uranus, and Neptune when it formed a semisquare or sesquisquare to each planet.

The lunar phase chart was activated by transiting Venus at 25 Aries in opposition to lunar phase Neptune, and the transiting Moon aligned with the lunar phase meridian and Venus and Uranus. Transiting planets in aspect to each other performed a function as well: Moon semisquare Sun (set off solar aspects), Venus trine Mars, and Mars sextile Neptune. In addition, transiting Mars was semisquare/sesquisquare the ingress horizon.

Transiting Mars at 25 Sagittarius was trine/sextile the lunar phase meridian and contraparallel the lunar phase Ascendant, and transiting Jupiter was parallel the lunar phase Ascendant. Transiting Venus and Jupiter were trine/sextile the lunar phase meridian.

In the ingress chart, transiting Mercury at 12 Pisces was trine/sextile the horizon and semisextile/inconjunct the meridian, and transiting Mars was semisquare/sesquisquare the horizon.

LAWRENCE, KANSAS, JULY 24, 1901

The Plains states experienced one of the hottest and most uncomfortable summers on record in 1901. In July, the thermometer reached 90 F every day of the month in Lawrence, Kansas. It soared to 100 F or higher on twenty-one days, including thirteen consecutive days from July 8th to the 25th. For nine days in a row, residents sweltered in 104 F heat and experienced the peak of 108 F on July 24.

The ingress chart (inner wheel of Temperature 6) had all the planetary configurations for a hot summer. Seven planets were in northern declination, and the Sun, Mercury, Venus, Jupiter, Saturn, Uranus, and Neptune were in high declination. But that only set the backdrop for what was to come with succeeding lunar phases throughout July.

Sun parallel Venus, contraparallel Jupiter, and conjunct Neptune all forecast coming heat, especially with the Sun semisquare/sesquisquare the meridian and inconjunct/semisextile the horizon. Neptune was inconjunct the Ascendant. Venus contraparallel Jupiter, in itself an aspect of heat, was intensified by the Sun-Venus parallel. The same was true of Venus semisextile Pluto. Pluto was square Mars, an aspect of blistering heat, and both planets aspected the meridian.

The planetary aspects and their connections to the longitude and latitude of Lawrence, while apparent, were not enough in themselves to indicate temperatures above 100 F for such an extended period of time. When combined with the lunar phase aspects, however, it was apparent that Lawrence would experience unprecedented heat. The solar eclipse of May 18, 1901, at 27 Taurus was trine/sextile the horizon.

The July 8, 1901, lunar phase chart got things simmering with Pluto trine/sextile the horizon and its square to the ingress meridian; Uranus opposed Pluto. The Sun had advanced to a semisextile with Pluto, fueling the Venus-Pluto aspect in both charts. Mars was square Neptune, and Venus had moved into sizzling Leo. Lunar phase Venus (heat) was parallel the ingress IC, and Mercury was contraparallel the ingress Ascendant. The lunar eclipse of May 3, 1901, at 12 Scorpio was square the lunar phase horizon.

By the next week, similar aspects were targeting Lawrence, only the chart positions had changed with the lunar phase of July 15, 1901. Lunar phase Mars was square the ingress Sun and square Neptune. Venus was semisquare Mars at 0 degrees declination, intensifying its characteristic heat. The ingress IC was parallel lunar phase Mercury and Venus, and the lunar phase Sun was contraparallel the ingress Ascendant.

The lunar phase of July 23, 1901 (outer wheel of Temperature 6), occurred the day before the thermometer peaked at 108 degrees. The Sun was semisextile Neptune and semisquare Pluto. Venus was semisquare Mars and sesquisquare Jupiter, and the added aspect of Mars square Jupiter pushed the temperature into an even higher range. Jupiter parallel Uranus indicated high pressure over the area. The Sun was trine and contraparallel the IC and semisextile the Ascendant, Mars was semisextile the Ascendant and Jupiter trined it, and Neptune was inconjunct the IC and sextile the Ascendant. Aspecting the meridian was the solar eclipse of May 18, 1901, at 27 Taurus, conjunct the Midheaven.

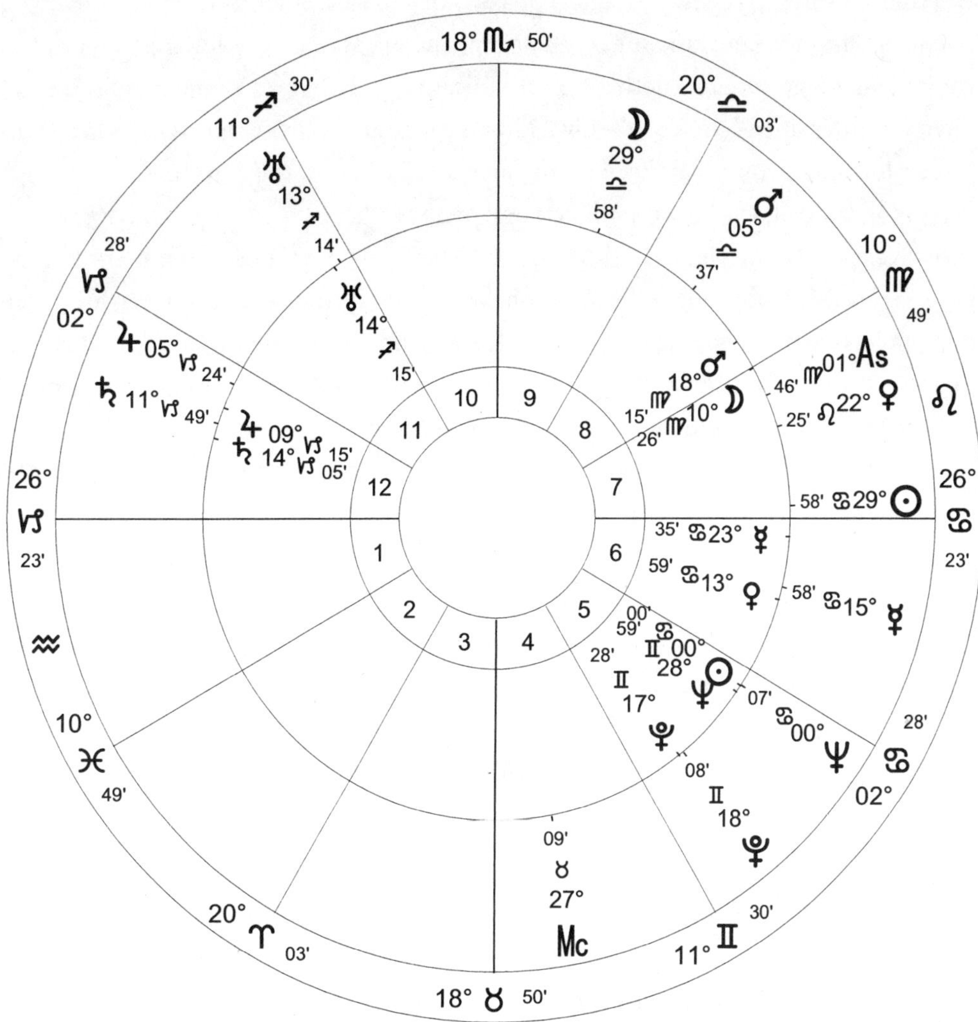

Temperature 6
Inner Wheel=*Summer Ingress / Natal Chart / June 21, 1901, Fri / 9:28 pm CST +6:00*
Lawrence, KS / 38°N58'18" 095°W14'06" / Geocentric / Tropical / Placidus / Mean Node

Outer Wheel=*Lunar Phase / Natal Chart / July 23, 1901, Tue / 7:58 am CST +6:00*
Lawrence, KS / 38°N58'18" 095°W14'06" / Geocentric / Tropical / Placidus / Mean Node

Aspects between the July 23 lunar phase chart and the ingress chart were many. Of major significance were lunar phase Pluto square ingress Mars and lunar phase Neptune conjunct the ingress Sun. Another aspect indicative of high heat, Venus sesquisquare Jupiter, was intensified (along with July 23 lunar phase and ingress aspects) by lunar phase Jupiter forming a contraparallel with the ingress Sun, Venus, and Neptune.

Important contacts between planets and angles in both charts targeted Lawrence for excessive heat. Lunar phase (and transiting) Mars was sesquisquare the ingress IC, ingress Venus was sesquisquare the lunar phase IC and semisquare the lunar phase Ascendant, ingress Sun was sextile the lunar phase IC, and lunar phase Venus was square the ingress meridian. The lunar phase Sun was conjunct and parallel the ingress Descendant, and the lunar phase meridian was sextile/trine the ingress horizon.

The ingress chart was activated on July 24 by these transits: Sun at 0 Leo semisextile Sun, semisquare Pluto and Mars, and sesquisquare Uranus; transiting Venus at 23 Leo sesquisquare Jupiter and parallel Pluto; Jupiter contraparallel Venus and the Sun; and Mars contraparallel Moon. The transiting Moon in Sagittarius/Scorpio first formed a trine to ingress Venus and a sesquisquare to ingress Neptune and then made a conjunction to the ingress Midheaven.

Cold Wave of 1899

Temperatures plummeted to below 0 F as far south as Louisiana and the Florida Panhandle during the mid-February 1899 cold wave, when an all-time low temperature was recorded in Milligan, Ohio: -39 F. This cold wave was preceded by another at the end of January/early February. Included in the February 1899 *Monthly Weather Review*, published by the American Meteorological Society, was this comment:

These cold waves established many new landmarks for future reference, whether we consider the instrumental reading or the physical phenomena resulting from the cold. The most striking of the latter perhaps was the flow of ice down the Mississippi River on the 17th, past New Orleans and into the Gulf of Mexico, an event never before witnessed within the memory of man. Ice an inch thick formed at the mouth of the Mississippi in East and Garden Island bays [Louisiana], and the temperature fell to 10 degrees F on the 13th. The loss of human life, from January 29th to February 13th, by freezing and avalanches (in Colorado) as near as can be ascertained was 105 persons.

Two inner planets, Mercury and Mars, were retrograde at the December 21, 1898, winter ingress (inner wheel of Temperature 7), and Venus was stationary direct. Retrograde and stationary planets often indicate extreme weather and characteristics that are opposite the norm; for example, retrograde Mars can indicate cold weather rather than warm.

On the date of the ingress in Milligan, Ohio; Tallahassee, Florida; and Minden, Lousiana, the Sun was conjunct retrograde Mercury in cold Capricorn, stationary direct Venus was conjunct Uranus (cold), and retrograde Mars was trine Uranus (cold) and square Jupiter (cold). There was also a long-term cold weather aspect: Saturn opposition Pluto. Six planets were in southern declination. Saturn was parallel Uranus, and both planets were contraparallel Neptune; all three were in declination with Mercury contraparallel Mars.

The lunar phase (a new Moon) in effect for the mid-February cold wave occurred on February 10 (outer wheel of Temperature 7). The new Moon in icy Aquarius was sextile Saturn (cold) and inconjunct retrograde Mars (cold). Mercury was square Jupiter in Aquarius and Scorpio, both cold signs, and sextile Uranus (cold). Mercury was approaching a trine/sextile to the ingress Pluto-Saturn opposition that would be exact a few days later. Retrograde Mars was sesquisquare Uranus (cold). A new outer-planet aspect had formed by the date of the lunar phase: Saturn opposition Neptune (cold, heavy precipitation). (Note: During this lunar phase period, the Washington, DC, area experienced a blizzard, with snow accumulation of 25–30 inches.)

There was a massive high-pressure system centered north of Montana on February 10 that had moved south by the morning of the 11th, and then developed into a closed high (extreme cold) that extended from Montana to the Dakotas and south into the central Plains by evening. On the 12th, the closed high moved farther south, centered in the Plains on the north and south Texas on the south. On February 13, the closed high moved eastward, with its southern edge over the Florida Panhandle. Note the strong presence of Uranus in both the ingress and lunar phase charts, and, to a lesser extent, Jupiter, both planets of high pressure.

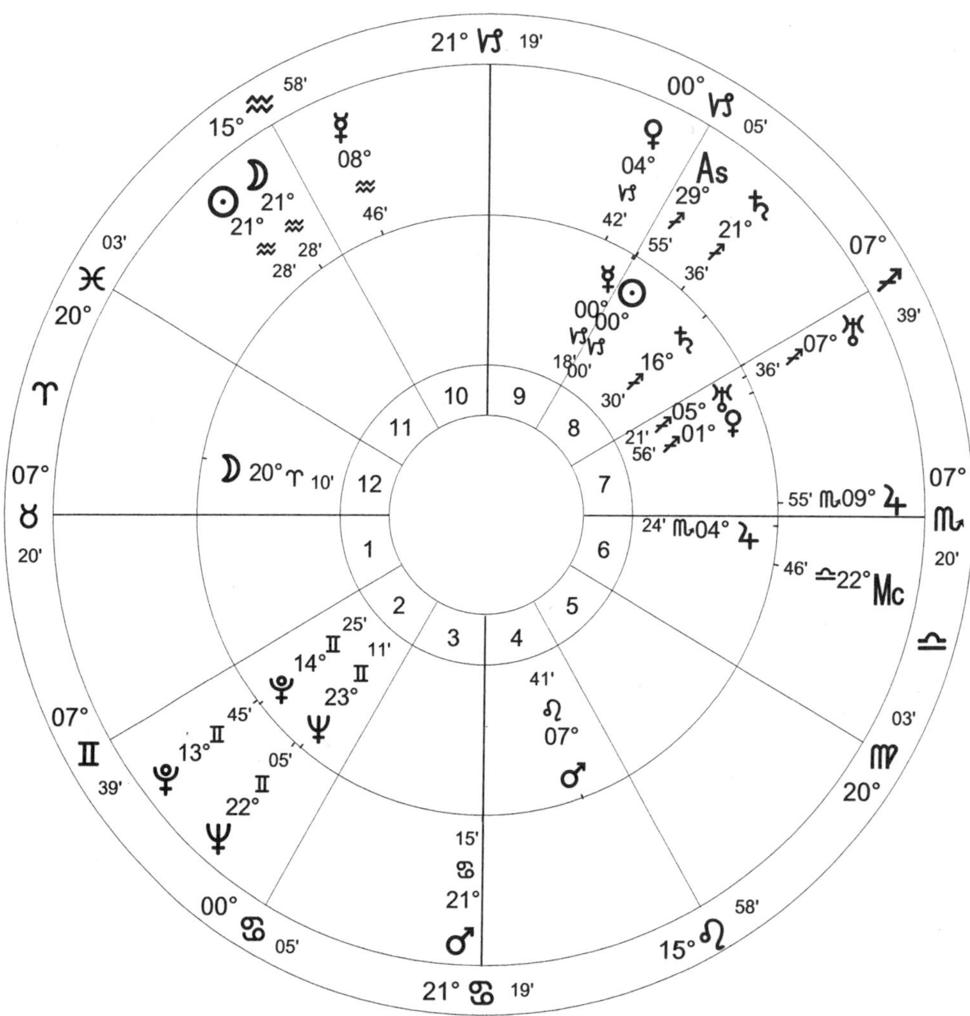

Temperature 7

Inner Wheel=*Winter Ingress / Natal Chart / December 21, 1898 NS, Wed / 12:59:11 pm CST +6:00*
Milligan, OH / 39°N44'25" 082°W06'09" / Geocentric / Tropical / Placidus / Mean Node

Outer Wheel=*Lunar Phase / Natal Chart / February 10, 1899 NS, Fri / 3:31:43 am CST +6:00*
Milligan, OH / 39°N44'25" 082°W06'09" / Geocentric / Tropical / Placidus / Mean Node

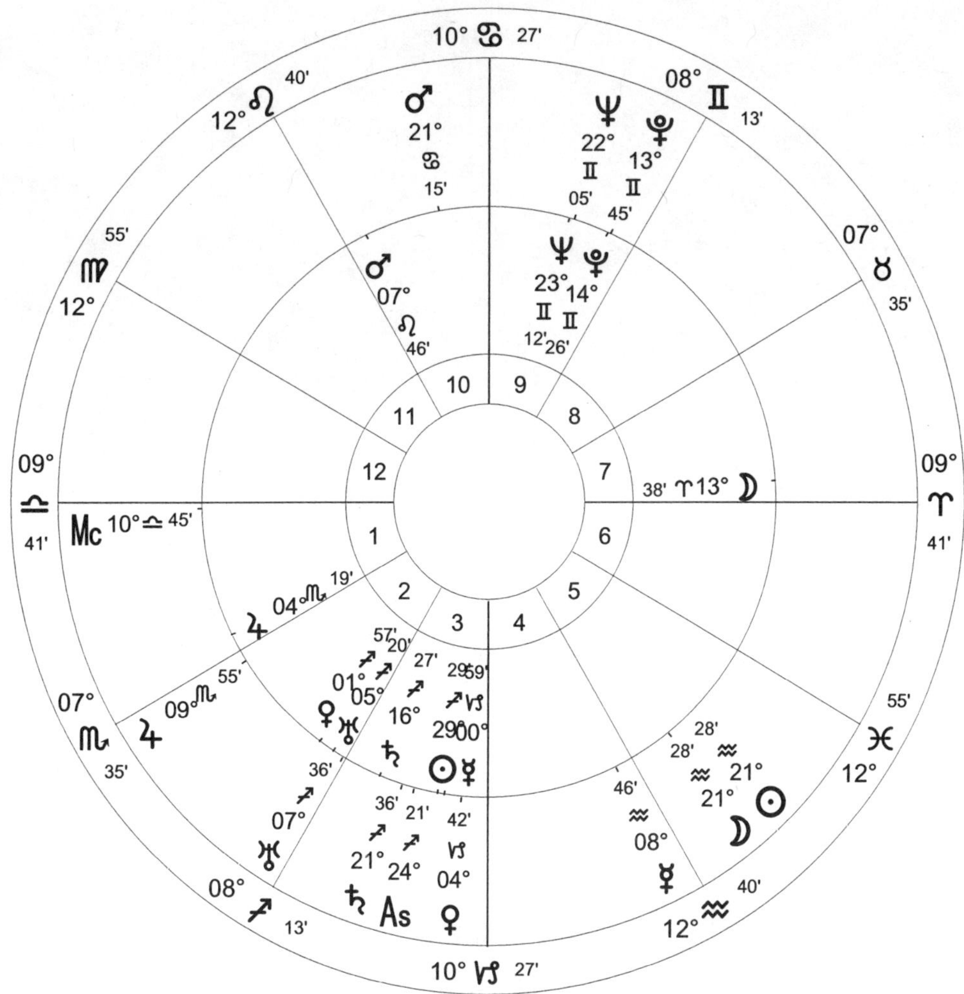

Temperature 8

Inner Wheel=*Winter Ingress / Natal Chart / December 21, 1898 NS, Wed / 0:59:11 am CST +6:00*
Minden, LA / 32°N36'55" 093°W17'12" / Geocentric / Tropical / Placidus / Mean Node

Outer Wheel=*Lunar Phase / Natal Chart / February 10, 1899 NS, Fri / 3:31:43 am CST +6:00*
Minden, LA / 32°N36'55" 093°W17'12" / Geocentric / Tropical / Placidus / Mean Node

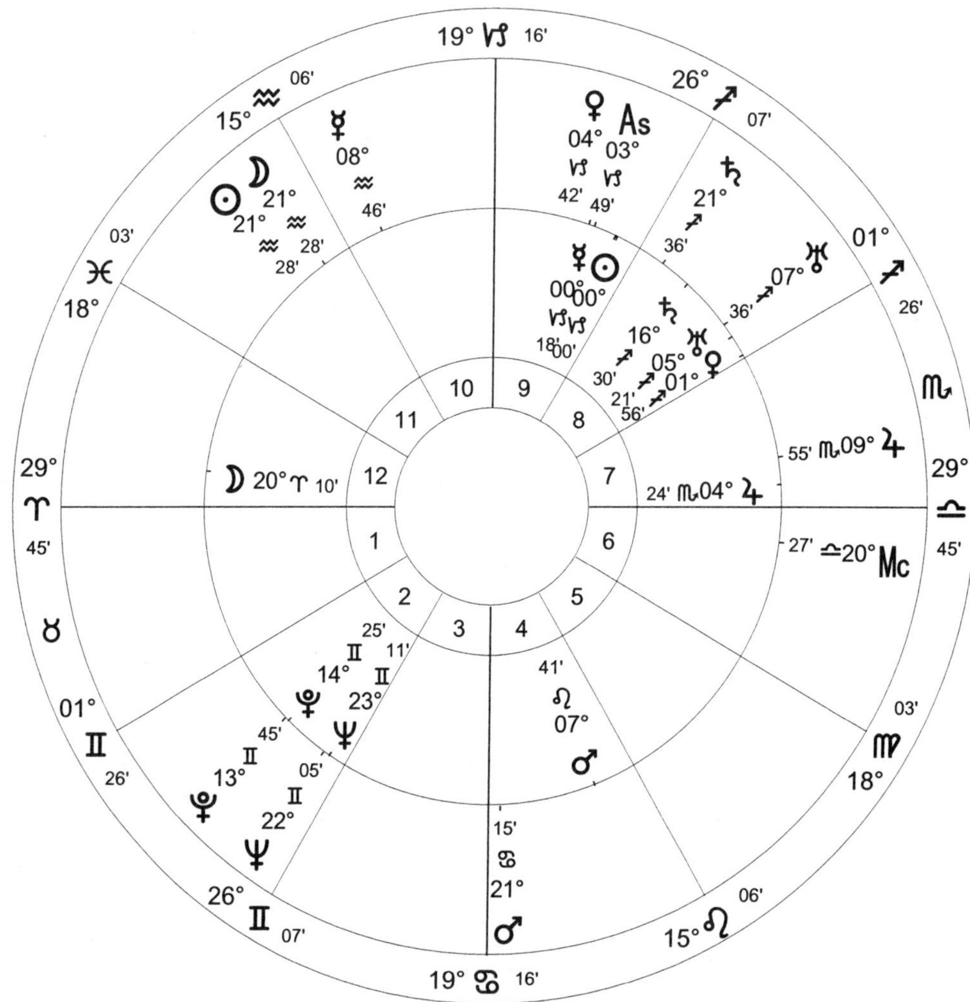

Temperature 9

INNER WHEEL=*Winter Ingress / Natal Chart / December 21, 1898 NS, Wed / 12:59:11 pm CST +6:00 Tallahassee, FL / 30°N26'17" 084°W16'51" / Geocentric / Tropical / Placidus / Mean Node*

OUTER WHEEL=*Lunar Phase / Natal Chart / February 10, 1899 NS, Fri / 3:31:43 am CST +6:00 Tallahassee, FL / 30°N26'17" 084°W16'51" / Geocentric / Tropical / Placidus / Mean Node*

The Ohio low temperature was caused by another meteorology factor: the 0°C isotherm. This tool is used by meteorologists to help determine temperature and what kind of precipitation an area will experience (rain, snow, sleet, etc.). Isotherms, which can be seen on meteorology maps, are a measure of the atmospheric temperature anywhere from the surface to about 1,000 mb (5,000 feet). Very cold temperatures are indicated at and north of the 0°C isotherm. On February 10, 1899, the 0°C isotherm extended from Maine, across the Great Lakes, into central Texas, and northwestward into Montana.

In Ohio, lunar phase retrograde Mars was conjunct/opposition the ingress meridian, activating the ingress Mars-Uranus trine (Temperature 7). The new Moon was trine/sextile the lunar phase meridian, as were lunar phase Saturn and ingress and lunar phase Neptune, activating the lunar phase Saturn-Neptune opposition. Lunar phase Uranus was semisquare/sesquisquare the lunar phase meridian. The lunar phase horizon aligned with the cold ingress retrograde Mercury-Sun conjunction. The solar eclipse of January 11, 1899, at 21 Capricorn was conjunct the ingress Midheaven.

In Louisiana, the lunar phase Saturn-Neptune opposition was conjunct the lunar phase horizon, and the new Moon was sextile/trine the same point (Temperature 8). The lunar phase and ingress meridians were aligned by square aspect. Lunar phase Mercury was opposition ingress Mars (cold), sextile lunar phase Uranus (cold), and semisquare/sesquisquare the lunar phase horizon. The transits on February 14 were the triggers, as Mercury in cold Aquarius formed a trine with ingress/lunar phase Pluto, a sextile with ingress Saturn, and a semisquare with the ingress retrograde Mercury-Sun conjunction in cold Capricorn. Transiting Venus in cold Capricorn was aligned with the ingress meridian, and the transiting Sun was semisquare/sesquisquare the same point.

Florida's low temperature, also on February 14, is reflected in transiting Mercury sextile/trine ingress/lunar phase Pluto and sextile ingress Saturn (cold), which was sextile/trine the ingress meridian. The meridians in the lunar phase and ingress charts were square, and lunar phase and transiting Mars were aligned with both meridians and sesquisquare the ingress Venus-Uranus conjunction (cold, high pressure). The new Moon also aspected both meridians, by semisextile and trine. The transiting Sun was sextile/trine the ingress horizon.

OMAHA, NEBRASKA, JANUARY–FEBRUARY 1936

During one of the coldest winters in Omaha's history, the average temperature from January 18 to February 19, 1936, averaged -2.8 F. This was the same year that the Great Plains experienced the highest temperatures and worst drought of the 1930s (see chapter 5).

The winter ingress chart (inner wheel of Temperature 10) is an excellent example of Saturn in aspect with Uranus, possibly the coldest long-term combination; the sextile indicates extreme cold. In the ingress chart, this aspect was intensified because the two planets were also contraparallel. In Omaha, as in area locations, both Saturn and Uranus aspected both the horizon and meridian through a trine/sextile to the meridian and semisextile/inconjunct to the horizon. Without a deeper study of the ingress chart, however, it would be difficult to forecast the extended cold wave based solely on this single aspect. By the time the bitter temperatures arrived in January, and throughout the period, Saturn and Uranus were no longer in aspect.

The ingress Sun was conjunct and parallel the Midheaven and trine Uranus, another indicator of cold weather, and one that activated the Saturn-Uranus sextile. The Uranus-Neptune configuration of unusual weather was present (a sesquisquare), and seven of the ten ingress planets were in southern declination (cold). The solar eclipse of July 30, 1935, at 8 Leo was trine/sextile the horizon, and the solar eclipse of December 25, 1935, at 3 Capricorn, three days after the ingress, was conjunction/opposition the ingress meridian.

The month-long deep freeze began two days after the January 16, 1936, lunar phase (outer wheel of Temperature 10), which featured Mercury in cold Aquarius conjunct and parallel the Midheaven. Venus was sesquisquare Uranus (overcast, cold), Mars was sextile and contraparallel Uranus (overcast, cold), and Jupiter sesquisquared Uranus (intense cold front). The Sun-Saturn semisquare was another indicator of cold temperatures.

Saturn and Uranus again aspected the horizon and meridian: Saturn inconjunct/semisextile the meridian and square the horizon, and Uranus semisextile/inconjunct the horizon. The lunar eclipse of January 8, 1936, at 17 Cancer was semisquare/sesquisquare the horizon, and seven planets were in southern declination.

Aspects between the ingress chart and the January 16 lunar phase chart targeted that week as the first of the intensely cold weather. The Ascendants were sextile, and ingress Mars (in Aquarius and sextile Jupiter) was conjunct the lunar phase Midheaven, indicating cold Arctic air pushed to the south. Ingress Jupiter was contraparallel the lunar phase IC.

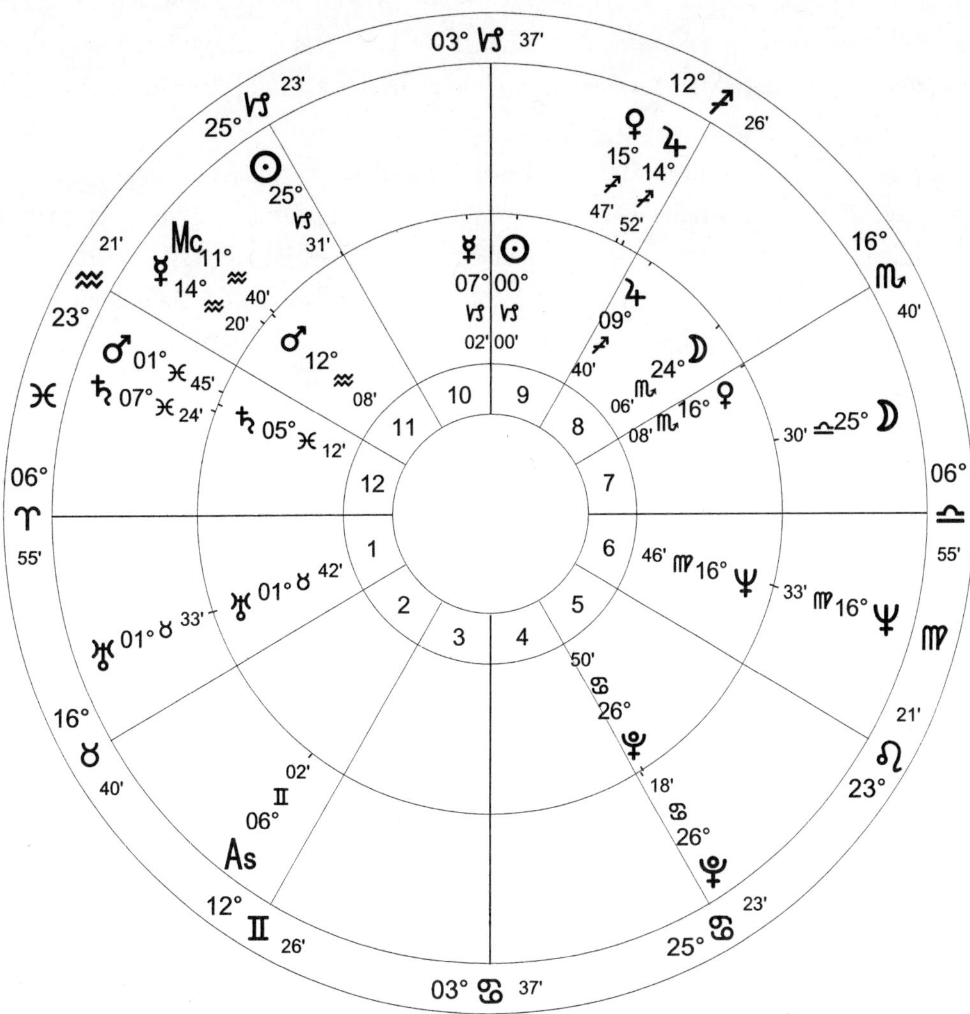

Temperature 10
Inner Wheel=*Winter Ingress / Natal Chart / December 22, 1935, Sun / 12:38 pm CST +6:00*
Omaha, NE / 41°N15'31" 095°W56'15" / Geocentric / Tropical / Placidus / Mean Node

Outer Wheel=*Lunar Phase / Natal Chart / January 16, 1936, Thu / 1:40 pm CST +6:00*
Omaha, NE / 41°N15'31" 095°W56'15" / Geocentric / Tropical / Placidus / Mean Node

Other important aspects between the two charts were lunar phase Jupiter sesquisquare ingress Uranus (intense cold front), lunar phase Venus sesquisquare ingress Uranus (overcast, cold), lunar phase Mars sextile ingress Uranus (overcast, colder), and lunar phase Saturn square ingress Jupiter (colder).

On January 18, transiting Mars at 3 Pisces formed an exact trine/sextile with the ingress meridian and a parallel with ingress Saturn in its approach to a conjunction with the same planet, while transiting Saturn at 7 Pisces was sextile ingress Mercury (cold). Transiting Mercury at 16 Aquarius was sesquisquare/semisquare the ingress meridian. The transiting Moon in Scorpio was square the lunar phase meridian and semisquare the ingress Sun, setting off the ingress Sun-Uranus trine. The transiting Moon also formed a semisextile to the Venus-Jupiter conjunction in the lunar phase chart (which was sesquisquare lunar phase Uranus) and a square to lunar phase Mercury, while the transiting Sun at 27 Capricorn was opposition ingress Pluto and parallel lunar phase Venus. Transiting Jupiter was contraparallel the lunar phase Ascendant. Two other factors added considerable thrust to the already apparent aspects: transiting Uranus turned stationary direct January 11, and transiting Mercury would form its retrograde station on January 23, the day before the next lunar phase.

In the lunar phase chart of January 24, retrograde Mercury in Aquarius indicated icy temperatures, as did Mars conjunct Saturn (cold). Uranus was aligned with the horizon and square the Sun and Moon. Seven planets were in southern declination.

In the lunar phase chart of January 30, when there were six planets in southern declination, the scenario was much the same, but more intense. Retrograde Mercury was conjunct the Sun in Aquarius, with both planets aligned with the horizon, while Uranus was conjunct and parallel the Midheaven. Mars, in a separating conjunction with Saturn, was approaching a semisquare to Uranus.

The icy blast continued into the next month with the lunar phase of February 7, 1936, and six planets in southern declination. In that chart, Uranus was in an almost exact conjunction with the IC, and retrograde Mercury, still in Aquarius, was square Uranus and the meridian. Saturn, which had formed an inconjunct/semisextile to the horizon in the January 30 chart, was sextile/trine the horizon in the February 7 chart.

When Mercury turned direct on February 13, residents of Omaha had only one more lunar phase to go before the bitterly cold temperatures would begin to lift. At the lunar phase of February 15, with seven planets in southern declination, Uranus was again prominent, this time in a nearly exact square with stationary direct Mercury. Jupiter, contraparallel the IC, continued its sesquisquare to Uranus, and Saturn again aspected the meridian and horizon. Lunar phase Mercury was inconjunct/semisextile and Uranus sextile the ingress meridian.

CHICAGO, ILLINOIS, JANUARY 10, 1982

Chicago, noted for its frigid January temperatures, lived up to its reputation on January 10, 1982, when the low was -26 F and winds blew at 25 mph. The combination resulted in a wind chill of -58 F. The icy temperatures on January 10 were only a single incident in what was one of the most severe winters on record for Chicago and Illinois. There were eighteen major winter storms, and the winter season was unusually long, extending from November through April.

At the winter ingress (inner wheel of Temperature 11), there were several key factors that indicated a cold winter: Sun semisextile Uranus is a cold wave configuration, as are Mars sextile Uranus, Jupiter semisextile Uranus, Saturn sextile Neptune, and Saturn contraparallel Pluto. Note also that the planetary configuration of prolonged cold was present in a semisquare: Saturn and Uranus. Mercury was prominent in its conjunction with the Descendant, the Sun was contraparallel the Ascendant, and Pluto was semisquare and parallel the IC. Venus, at its coldest in Aquarius, was inconjunct/semisextile the horizon. Eight planets were in southern declination.

The many lunar phase aspects (and lunar phase–ingress aspects) involving Uranus reflect the extreme Arctic high-pressure system and the blast of cold air that was channeled into the area; a strong low-pressure to the east kept the cold air contained in a much narrower area, preventing it from dispersing over a wider one. The barometer rose to an incredibly high 1057 mb. (By comparison, the highest pressure ever recorded in the continental United States was 1064 mb at Miles City, Montana, on December 24, 1983.) The Saturn and Venus aspects (and the approaching semisquare of the Mercury-Venus conjunction) were precursors of the storm of January 12–13.

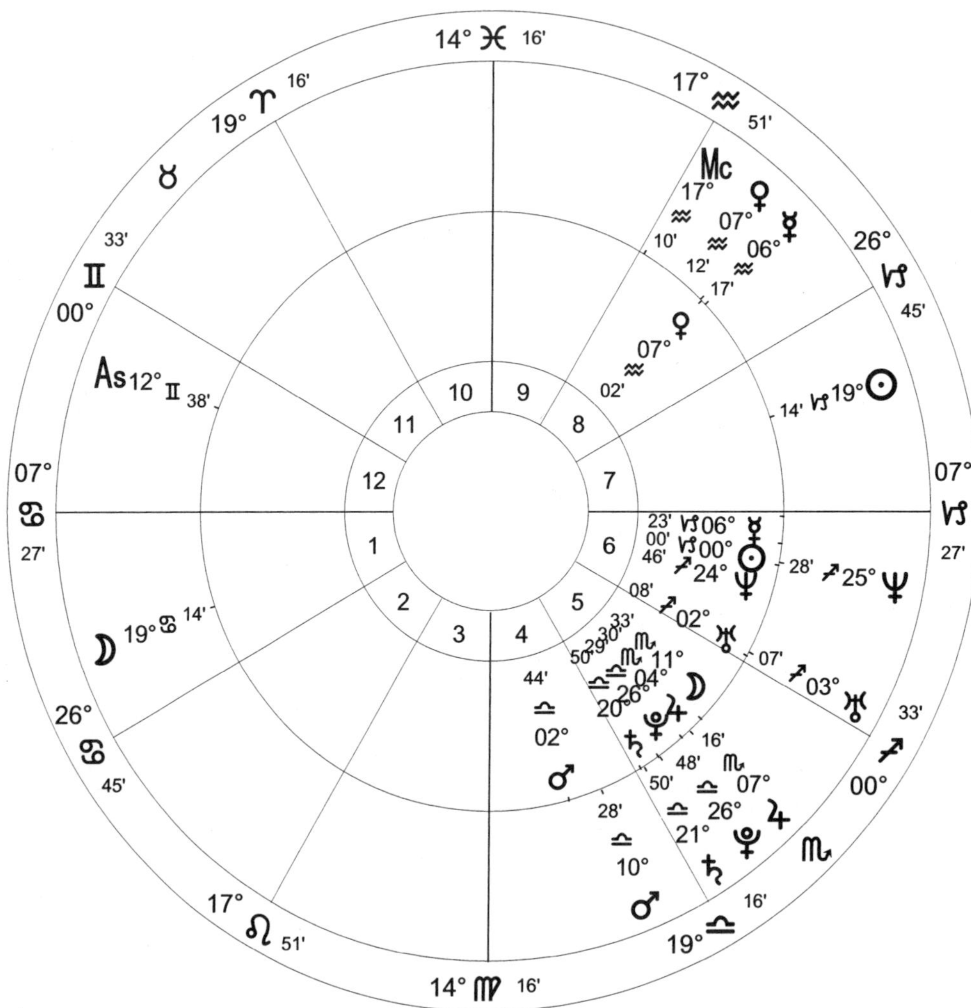

Temperature 11
Inner Wheel=*Winter Ingress / Natal Chart / December 21, 1981, Mon / 4:51 pm CST +6:00*
Chicago, IL / 41°N51' 087°W39' / Geocentric / Tropical / Placidus / Mean Node

Outer Wheel=*Lunar Phase / Natal Chart / January 9, 1982, Sat / 1:53 pm CST +6:00*
Chicago, IL / 41°N51' 087°W39' / Geocentric / Tropical / Placidus / Mean Node

By the time of the January 9, 1982, lunar phase (outer wheel of Temperature 11), Venus, still in Aquarius, had turned retrograde, aggravating its tendency toward cold. Its conjunction with Mercury, with both planets square Jupiter, targeted Chicago through Venus's contraparallel to the IC. More intensity was indicated by Mercury's sextile and parallel to Uranus. Other indicators of cold weather were the Sun square Saturn, semisquare Uranus, inconjunct/semisextile the meridian, and aligned with the horizon by declination, as well as the still-present Saturn-Uranus semisquare. The wind, which only aggravated the cold, was indicated by Pluto sesquisquare the lunar phase Ascendant, retrograde Venus conjunct Mercury (semisextile/inconjunct the ingress horizon) and trine Mars, and ingress Mars sesquisquare/semisquare the lunar phase horizon.

Lunar phase Venus formed a conjunction with ingress Venus, and both were in declination aspect with the lunar phase meridian. The lunar phase Sun was square Saturn in both charts. Further links between the two charts were the lunar phase horizon square the ingress meridian and the horizons of both charts parallel each other.

On January 10, 1982, the transiting Sun at 19 Capricorn was square ingress Saturn, and the conjunction of Venus to its ingress position was exact. Transiting Venus at 7 Aquarius was also semisextile ingress Mercury, transiting Mercury at 7 Aquarius was conjunct ingress and lunar phase Venus, and transiting Jupiter at 7 Scorpio was square ingress Venus—all demonstrating the chilling power of planets in Aquarius. Transiting Mercury was parallel ingress Uranus (descending cold air), transiting Saturn was in declination aspect with the ingress meridian, and transiting Mercury formed an inconjunct with the ingress Ascendant that morning; the Mercury aspects reflected windy conditions.

The transiting planets also impacted the lunar phase chart. Transiting Venus at 7 Aquarius formed a conjunction with lunar phase Mercury and was square lunar phase Jupiter, which also was square transiting Mercury at 7 Aquarius. The transiting Moon in early Leo was trine and contraparallel lunar phase Uranus and opposed lunar phase Venus and Mercury. It also formed a contraparallel with lunar phase Mercury.

Transiting aspects that day were similar: Mercury conjunct Venus at 7 Aquarius, and Mercury and Venus square Jupiter at 7 Scorpio. The triggering Moon in early Leo was also sesquisquare/semisquare the ingress meridian and semisquare/sesquisquare the lunar phase horizon.

CHAPTER 8

Hurricanes

Some of the worst weather-related disasters in history have occurred as a result of the destructive force of a hurricane, which comes from two factors: wind and water. Every fall, meteorologists watch storms developing near the Cape Verde Islands off the coast of Africa and then track their progress across the Atlantic Ocean to the East Coast of the United States. Hurricanes that strike the U.S. can also form in the Gulf of Mexico, the Caribbean Sea, or near Baja California. When hurricanes develop in the Indian Ocean, they are called cyclones; those that develop in the Pacific Ocean are called typhoons.

The evolution of a hurricane from tropical depression to tropical storm to hurricane as it gathers strength while traveling from the African coast to the U.S. coast can take a week or more—ample time to build into a mass of destruction and to warn coastal residents *if* meteorologists and computer modeling correctly predict *when* and *where* it will make landfall. Determining these three factors—if, when, and where—is an area where astrometeorology can be of tremendous value. Even if meteorology is used to forecast a hurricane, the astrometeorologist can determine exactly when and where the eye will cross land, thereby saving lives and property.

A hurricane is a massive collection of bands of thunderstorms spiraling counterclockwise toward the center (eye) and moving generally east to west/west to east or north to south/south to north over an ocean. Typically measuring 300 miles in circumference, with winds of 74 mph or more, a hurricane can produce waves in excess of fifty feet. Unlike with an extratropical cyclone (a major—usually winter—storm over land that has a counterclockwise movement and looks similar to a hurricane on satellite images), the center of a hurricane is warmer than the surrounding air.

Hurricanes form in the tropics because the climate provides the necessary ingredients: a wide expanse of warm water and warm, humid air. Warm water powers hurricanes, and inside a hurricane, humid air rises and condenses into drops that release heat. The heat warms the surrounding air, making it lighter. The lighter air rises, and as warm air rises, more flows in to replace it, creating wind. High pressure in the upper atmosphere helps pump away the rising air.

As the hurricane travels across the ocean, winds push the water toward its center, and as long as the hurricane remains offshore, this wall of water is mostly submerged. However, when the eye, which has lower atmospheric pressure, crosses land, the water wall rises and surges over the land, with the wind pushing it. This water surge often causes more damage to coastal areas than does the wind of a hurricane. It can be as much as twenty feet high, and is compounded by the six to twelve inches of rain that generally accompany the storm's arrival on land. The heaviest precipitation and highest winds in a hurricane are on the right side, in the right front quadrant of the storm, where tornadoes also often occur once the storm makes landfall.

Planetary signatures of a hurricane are hard and neutral aspects between Mercury and Uranus, Mercury and Neptune, Mars and Uranus, and Mars and Neptune. Mercury or Mars represents the wind, Uranus represents the upper atmosphere high pressure necessary to keep the wind/water hurricane engine running, and Neptune represents counterclockwise winds and low pressure. With strong Saturn aspects, the pressure in the eye can be much lower. Note that the key ingredients of a hurricane are the same as those for a tornado, indicating the counterclockwise air movement typical of both phenomena.

HURRICANE SANDY, OCTOBER 29, 2012

Hurricane Sandy, which evolved into Superstorm Sandy, affected twenty-four American states, including the entire East Coast from Florida to Maine, and caused particularly severe damage in New York and New Jersey, where it made landfall at about 8:00 pm with 80 mph winds. Its storm surge flooded streets and subway tunnels, and many of New

Jersey's beaches shrunk by thirty to forty feet. Power outages affected 10 million people, and numerous fires ignited as a result of ruptured natural gas mains. Total damage was more than $65 billion, making it the second-costliest storm in U.S. history. At least 286 people died. Ingress and lunar phase charts for Atlantic City, New Jersey, are used here, as this city was close to where Sandy made landfall.

The September 22, 2012, autumn ingress chart (Hurricanes 1) showed strong potential for heavy precipitation, flooding, and a hurricane: Neptune conjunct the IC, which was compounded by its placement in wet Pisces; this is an example of a planetary conjunction to the IC that shows the greatest impact to the north. With this planet so prominent in the New Jersey/New York area at the ingress, a hurricane was almost guaranteed. Mars (wind) was conjunct the Ascendant and sesquisquare Uranus, another hurricane aspect that also indicates wind and stormy conditions. There was a third hurricane aspect at the ingress: Mercury opposition Uranus (hurricane, wind). Venus was square Mars and the Ascendant, another storm and precipitation aspect, and the area was also prime for a low-pressure system because of ingress Saturn sextile/trine the meridian.

Sandy began on October 22, 2012, as a low-pressure system in the Caribbean Sea, as the transiting Sun formed a conjunction with transiting Saturn conjunct the IC in the Caribbean (Hurricanes 2). Both planets were sesquisquare Jupiter, an aspect of storms and low pressure, and Saturn was trine Neptune (precipitation). The tropical disturbance quickly strengthened and was upgraded to Tropical Storm Sandy, gradually intensifying as it moved northward. Sandy became a hurricane on October 24, and crossed Jamaica, Cuba, the Dominican Republic, Puerto Rico, and the Bahamas on its path toward the United States, during which it was upgraded and downgraded several times.

Sandy's progress through the Caribbean and northward along the American East Coast occurred while the lunar phase of October 21, 2012, was in effect. Even though its impact on the northeastern United States the night of October 29, 2012, was devastating, by that time Sandy was no longer a tropical storm. It briefly reintensified to a Category 2 storm that morning, and then diminished as it approached the New Jersey coast after the new lunar phase went into effect that afternoon.

As obvious as the hurricane potential is in the ingress chart, it is less so in the lunar phase chart for Atlantic City (Hurricanes 3) and the comparison between the two because there are no specific hurricane signatures. Nevertheless, there were significant aspects.

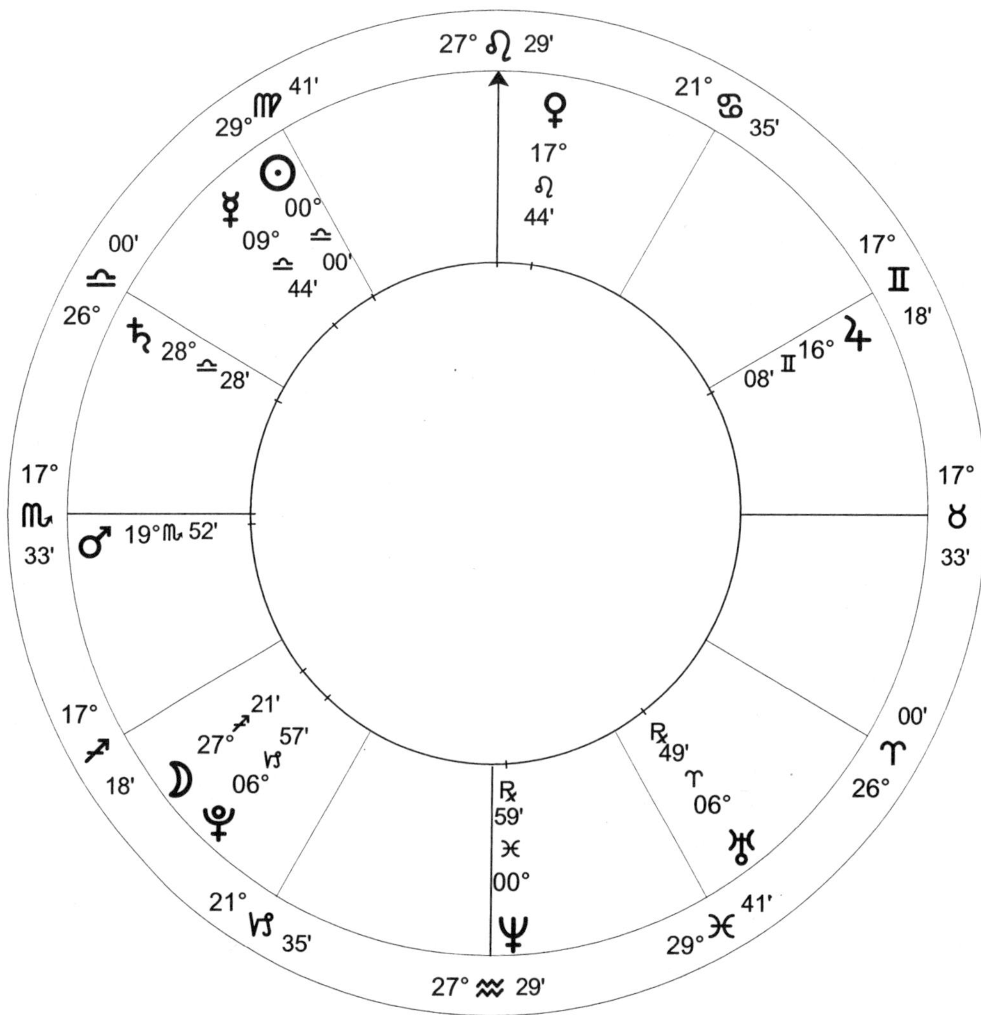

Hurricanes 1: Autumn Ingress / Natal Chart / September 22, 2012, Sat / 10:48:59 am EDT +4:00
Atlantic City, NJ / 39°N21'51" 074°W25'24" / Geocentric / Tropical/ Placidus / Mean Node

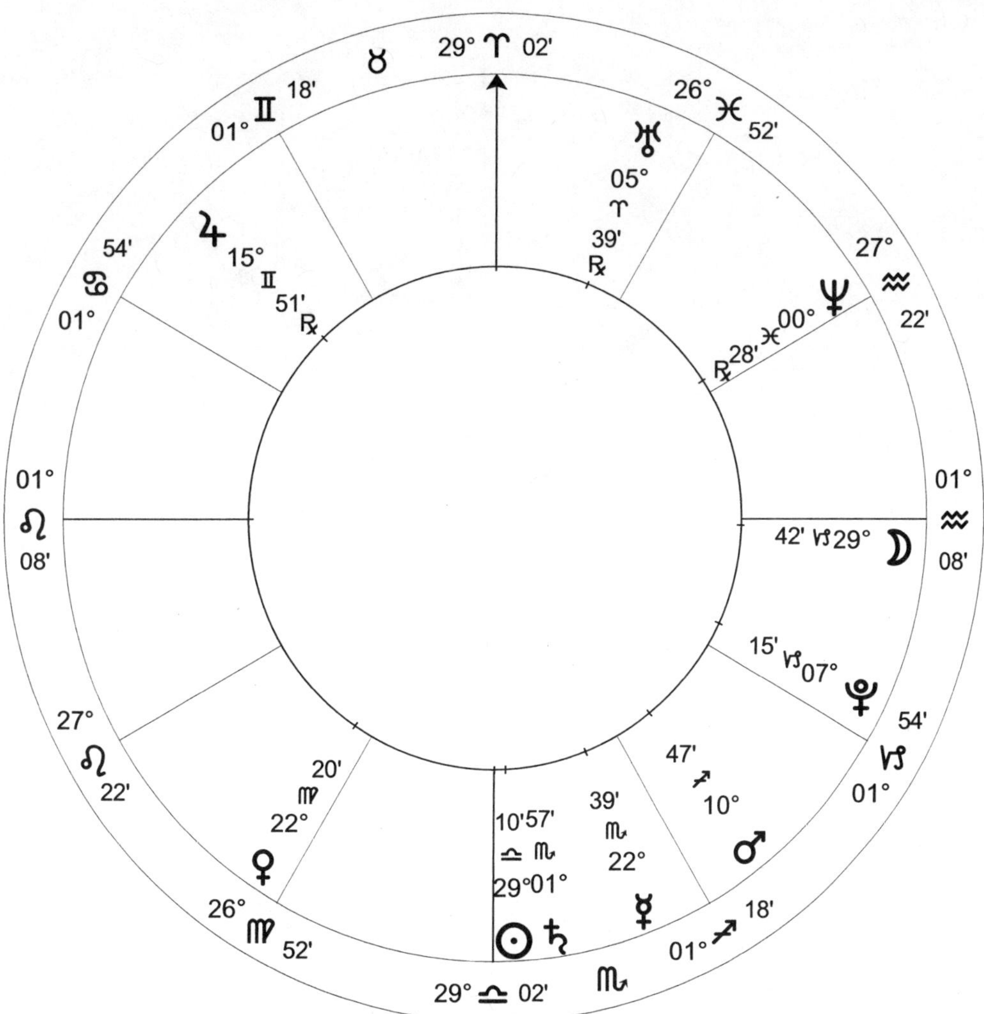

Hurricanes 2: Lunar Phase / Natal Chart / October 21, 2012, Sun / 11:31:58 pm EST +5:00
Caribbean Sea / 17°N00' 072°W00' / Geocentric / Tropical / Placidus / Mean Node

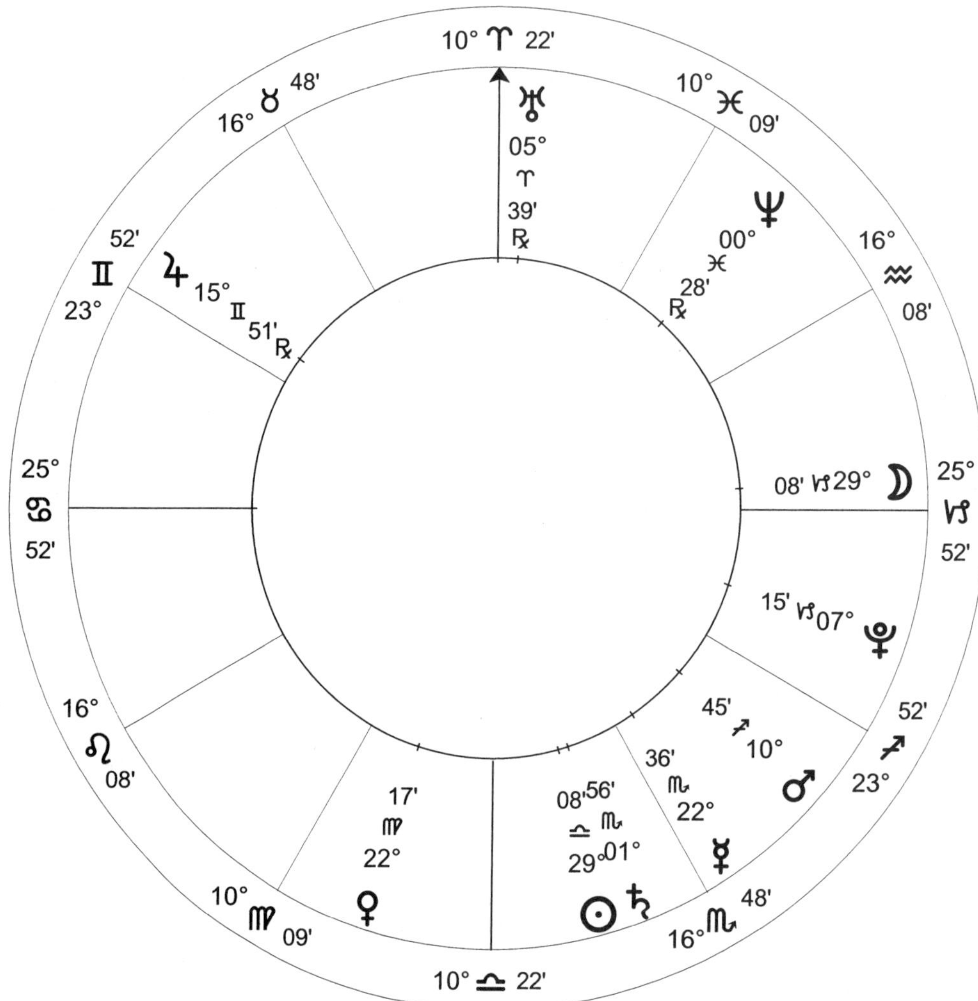

Hurricanes 3: Lunar Phase / Natal Chart / October 21, 2012, Sun / 11:31:58 pm EDT +4:00
Atlantic City, NJ / 39°N21'51" 074°W25'24" / Geocentric / Tropical / Placidus / Mean Node

The lunar phase chart of October 21, 2012, for Atlantic City had a Uranus-Pluto square (extreme weather and high winds) in aspect to the meridian. Mars was semisextile Pluto and aspecting both the meridian (trine/sextile) and the horizon (sesquisquare/semisquare), bringing the Uranus-Pluto aspect into effect at that longitude and latitude. Jupiter was parallel the Ascendant and sesquisquare Saturn (storm, low pressure).

Ingress Mercury was conjunct the lunar phase IC (activating the ingress Mercury-Uranus opposition), the lunar phase horizon was square ingress Saturn, and the lunar phase Sun-Saturn conjunction was trine the ingress IC conjunct Neptune.

As the storm arrived in New Jersey, transiting Mercury was square the ingress IC-Neptune conjunction, activating the hurricane potential of the ingress chart. Transiting Mars was semisextile and transiting Venus semisquare the ingress Ascendant. These transits were the final indicators of where Sandy would make landfall in the northeastern United States.

However, Sandy was not finished after it made landfall. Instead, it became Superstorm Sandy, evolving from a tropical cyclone into an extratropical cyclone. Look again at the lunar phase chart for Atlantic City and notice Uranus in the ninth house. This represents a blocking high (high-pressure system) over the Atlantic Ocean that prevented Sandy from taking the usual hurricane path; most hurricanes arc to the right, missing the northeastern United States.

So Sandy had nowhere to go but farther onto land. And there it encountered a large, intense atmospheric trough (a low wavy loop that was the opposite of the high wavy loop of the high-pressure system over the Atlantic). Sandy essentially became trapped between the deep trough and the high-pressure system. Both of these systems are reflected in the October 29 lunar phase chart (Hurricanes 4). Jupiter to the left of the IC represents the high-pressure system, and Mercury to the left of the Midheaven and square Neptune represents the trough. Mercury also indicates the strong cold front associated with the trough.

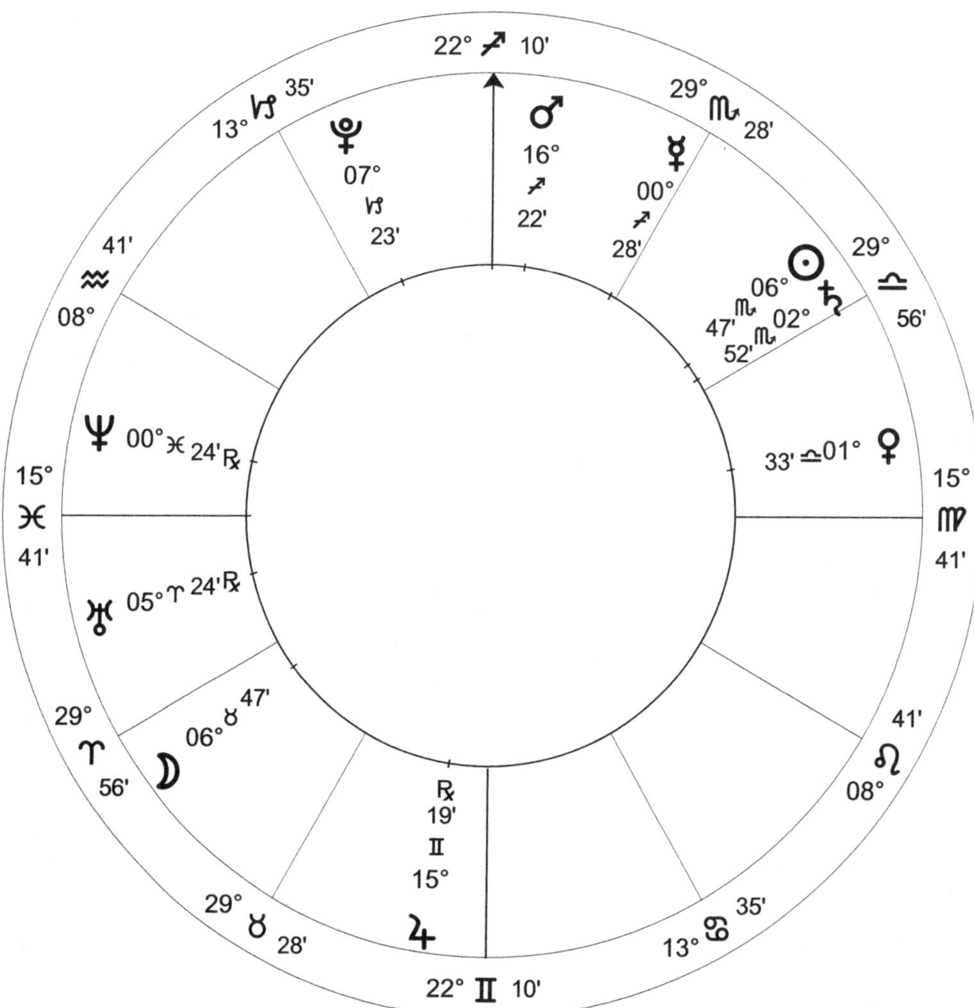

Hurricanes 4: Lunar Phase / Natal Chart / October 29, 2012, Mon / 3:49:26 pm EDT +4:00
Atlantic City, NJ / 39°N21'51" 074°W25'24" / Geocentric / Tropical / Placidus / Mean Node

Sandy was able to transition from tropical to extratropical because of an influx of warm, tropical air (lunar phase Venus semisquare/sesquisquare the ingress horizon) and cold air (lunar phase Uranus sesquisquare/semisquare the ingress horizon). This was the only time in recorded history that a hurricane transitioned into a hybrid storm. Its effects were in evidence as far west as the Great Lakes, creating waves in some locations as high as twenty-three feet, and three feet of heavy, wet snow accumulated in West Virginia and North Carolina.

HURRICANE CHARLEY, AUGUST 13, 2004

Charley, classified as Category 4, was the strongest hurricane to hit Southwest Florida since Hurricane Donna in 1960. It made landfall near Port Charlotte, which, along with other nearby cities, was isolated for nearly two days because downed trees, power poles, power lines, and debris filled the streets. Buildings collapsed under wind gusts as high as 147 mph. Damage was more than $13 billion, and thirty-five people died as a result of the storm.

The June 20, 2004, summer ingress chart (inner wheel of Hurricanes 5) had no major indications of hurricane potential unless we look a bit closer. Mercury was sesquisquare Neptune (hurricane), with Neptune sesquisquare/semisquare the meridian and semi-sextile/inconjunct the horizon. What added to the potential of the weaker semisquare was retrograde Venus square Jupiter and trine Neptune (heavy precipitation). Pluto (wind) was sextile/trine the meridian, and although Uranus was sextile/trine the horizon and sesquisquare/semisquare the meridian, it made no aspects to other planets.

This condition changed at the August 7, 2004, lunar phase (middle wheel of Hurricanes 5), where stationary retrograde Mercury was opposition both lunar phase and ingress Uranus, which was parallel the lunar phase Midheaven. By the time of the lunar phase, Saturn was sesquisquare Uranus (storm) and semisquare Mercury. Lunar phase Venus was sesquisquare Neptune (heavy precipitation).

There is an obvious reason why Charley targeted Port Charlotte: the signs on the lunar phase and ingress charts are identical and the degrees are very close, a factor often seen in major weather events. Add to that stationary retrograde Mercury square ingress retrograde Venus, with Mercury trine the ingress Ascendant, and lunar phase Saturn (low pressure) square the ingress meridian; lunar phase Mars was semisquare ingress Saturn (wind, storm).

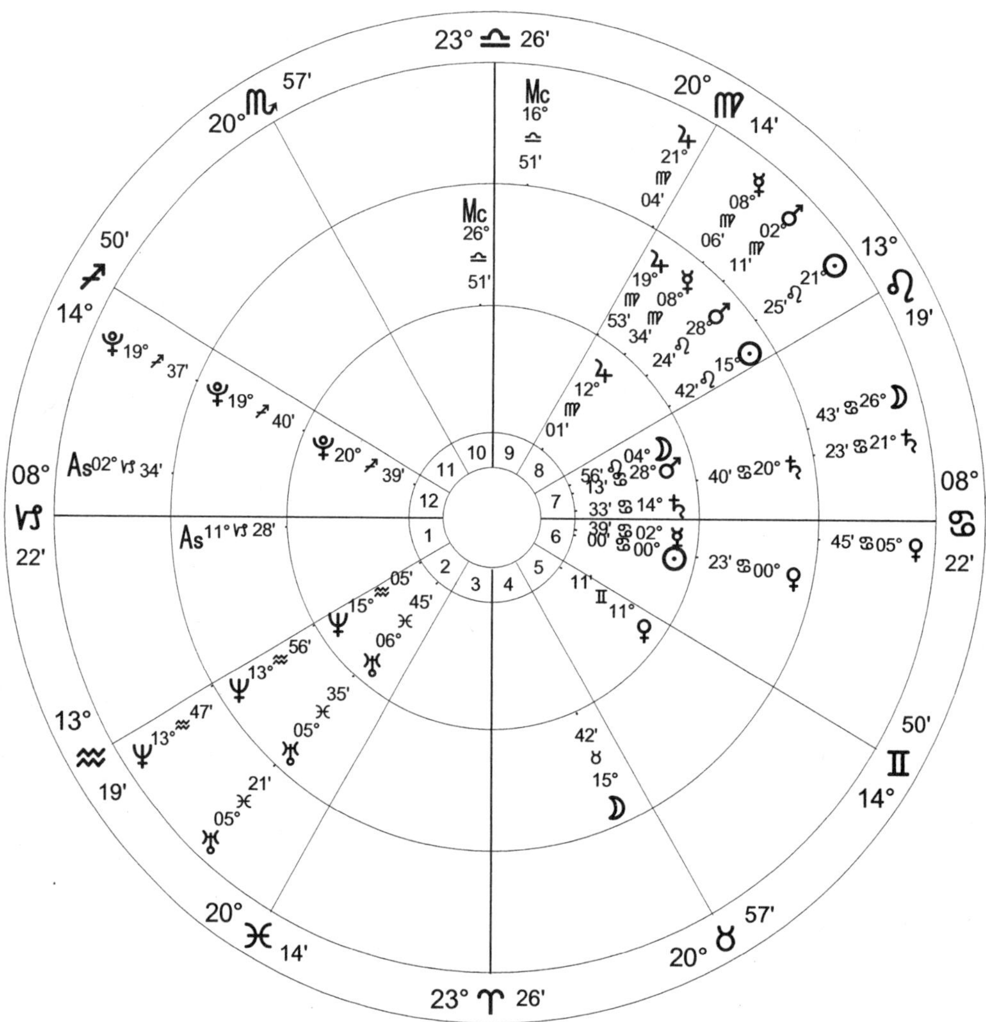

Hurricanes 5

Inner Wheel=*Summer Ingress / Natal Chart / June 20, 2004, Sun / 8:56:52 pm EDT +4:00 Port Charlotte, FL / 26°N58'33'' 082°W05'27'' / Geocentric / Tropical / Placidus / Mean Node*

Middle Wheel=*Lunar Phase / Natal Chart / August 7, 2004, Sat / 6:01 pm EDT +4:00 Port Charlotte, FL / 26°N58'33'' 082°W05'27'' / Geocentric / Tropical / Placidus / Mean Node*

Outer Wheel=*Hurricane Charley / Natal Chart / August 13, 2004, Fri / 5:00 pm EDT +4:00 Port Charlotte, FL / 26°N58'33'' 082°W05'27'' / Geocentric / Tropical / Placidus / Mean Node*

As Charley made landfall (outer wheel of Hurricanes 5), transiting Venus (precipitation) was opposition/conjunction the ingress horizon, and the transiting Sun was trine Pluto and the ingress IC (wind). Most telling of the power of the hurricane, however, was transiting Mars opposition ingress Uranus, which aspected both the meridian and horizon in the ingress chart.

HURRICANES FRANCES AND JEANNE, 2004

Two hurricanes hit Stuart, Florida, in 2004, one in August (summer ingress) and the second in September (autumn ingress). The two charts are obviously very different, so we'll look at the planetary indicators and positions that targeted this city twice during the same year.

The charts for Hurricane Jeanne bring up a dilemma that you'll see from time to time. The autumn ingress occurred on September 22, but the lunar phase nearest the hurricane was September 21. The summer ingress was still in effect on September 21, and the autumn ingress had yet to occur. In this situation, I usually use the ingress that is within a few days of the lunar phase, but you should experiment and draw your own conclusions. Remember this as you look at the charts for Hurricane Jeanne, because the lunar phase planets are behind their positions at the ingress a day later.

The June 20, 2004, summer ingress chart for Hurricane Frances (inner wheel of Hurricanes 6) is nearly identical to that for Hurricane Charley because these hurricanes were in the same season and made landfall in the same state. This brings up the question of why Frances targeted Stuart instead of Port Charlotte. The answer is in the August 29, 2004, lunar phase chart (middle wheel of Hurricanes 6), where Saturn (low pressure) was aligned with the meridian at 22 Cancer/Capricorn. In the lunar phase chart for Port Charlotte for the same date, the meridian is 20 Cancer/Capricorn. So Saturn was closer to the lunar phase meridian in Stuart. The horizon in the Port Charlotte lunar phase chart was 0 Taurus/Scorpio, and in the Stuart chart it was 2 Taurus/Scorpio. At the time of Hurricane Frances, transiting Mars (square Pluto, wind) was exactly sesquisquare the Stuart horizon, but in Port Charlotte it was 2 degrees from exact. The Mars and Saturn aspects thus targeted Stuart for Frances's landfall.

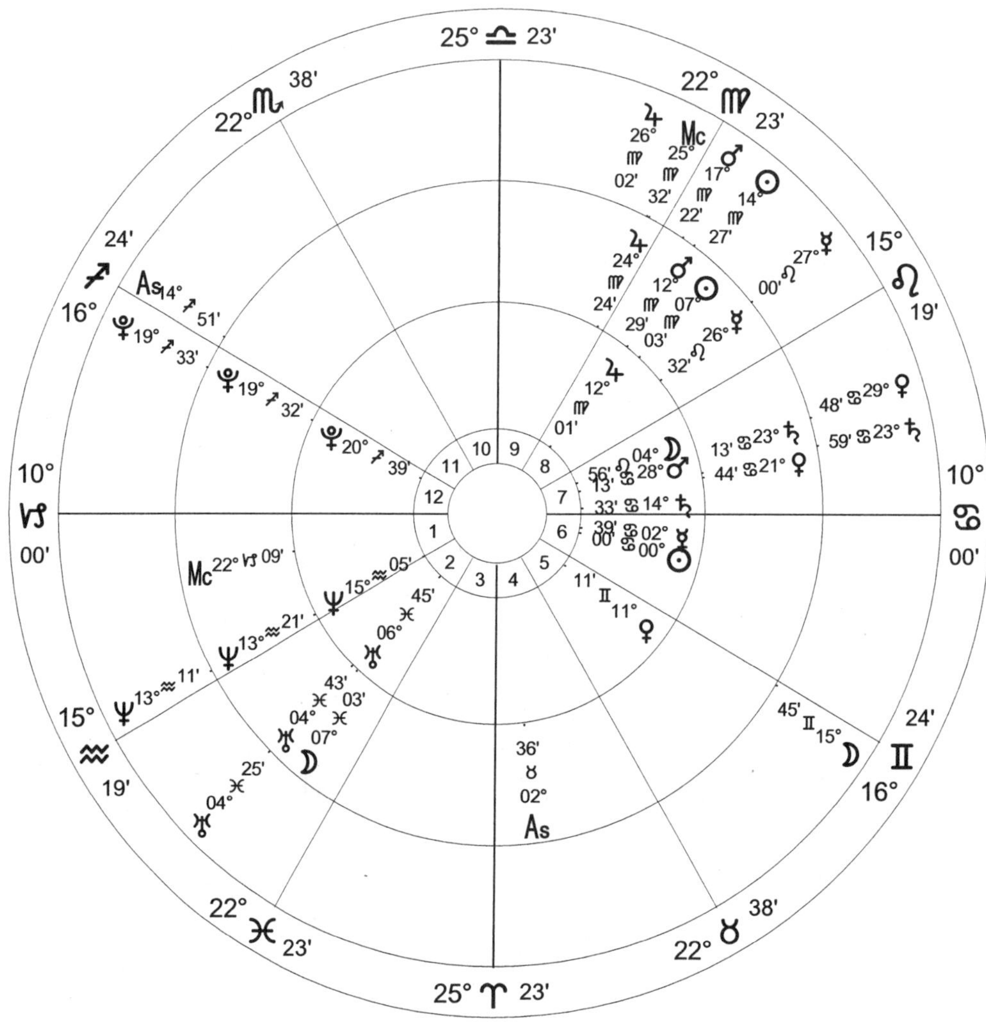

Hurricanes 6

INNER WHEEL=*Summer Ingress / Natal Chart / June 20, 2004, Sun / 8:56:52 pm EDT +4:00*
Stuart, FL / 27°N11'50'' 080°W15'11'' / Geocentric / Tropical / Placidus / Mean Node

MIDDLE WHEEL=*Lunar Phase / Natal Chart / August 29, 2004, Sun / 10:22:15 pm EDT +4:00*
Stuart, FL / 27°N11'50'' 080°W15'11'' / Geocentric / Tropical / Placidus / Mean Node

OUTER WHEEL=*Hurricane Frances / Natal Chart / September 6, 2004, Mon / 2:00 pm EDT +4:00*
Stuart, FL / 27°N11'50'' 080°W15'11'' / Geocentric / Tropical / Placidus / Mean Node

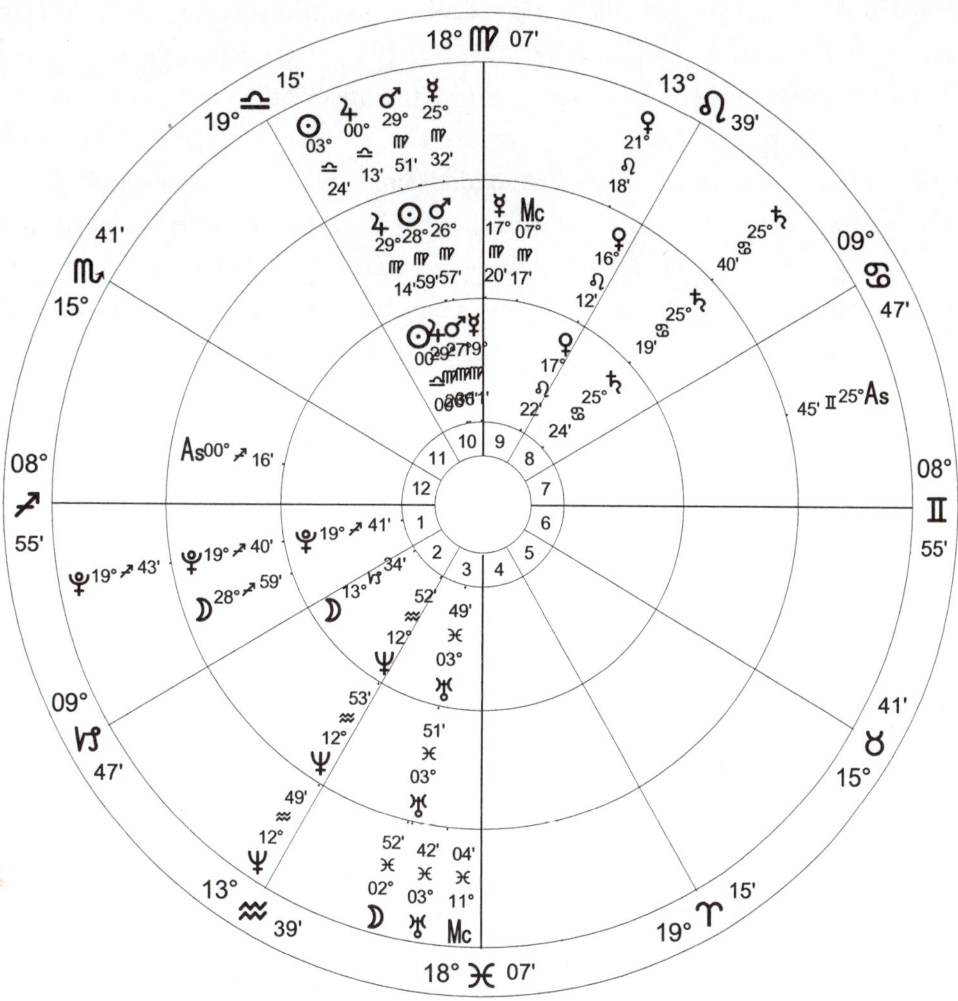

Hurricanes 7

INNER WHEEL=*Autumn Ingress / Natal Chart / September 22, 2004, Wed / 12:29:50 pm EDT +4:00*
Stuart, FL / 27°N11'50" 080°W15'11" / Geocentric / Tropical / Placidus / Mean Node

MIDDLE WHEEL=*Lunar Phase / Natal Chart / September 21, 2004, Tue / 11:53:34 am EDT +4:00*
Stuart, FL / 27°N11'50" 080°W15'11" / Geocentric / Tropical / Placidus / Mean Node

OUTER WHEEL=*Hurricane Jeanne / Natal Chart / September 25, 2004, Sat / 11:50 pm EDT +4:00*
Stuart, FL / 27°N11'50" 080°W15'11" / Geocentric / Tropical / Placidus / Mean Node

Now look at the ingress charts for Hurricane Jeanne. (Because they are only a day apart, the lunar phase and ingress charts are nearly identical except for the horizon and meridian.) The most obvious ingress aspect that made Stuart a target for the second time during the 2004 hurricane season was Mercury conjunct the Midheaven and square Pluto, which was also square the meridian (inner wheel of Hurricanes 7). When Hurricane Jeanne made landfall three days after the ingress (outer wheel of Hurricanes 7), transiting Mercury was conjunct lunar phase/ingress Mars (wind). Ingress Saturn (low pressure) was semisquare/sesquisquare the ingress meridian, and ingress Neptune, which was sesquisquare ingress Mars (hurricane, wind, low pressure), was sextile/trine the horizon.

Jeanne, a Category 3 storm, turned north and remained over land to the Florida-Georgia state line and eventually became an extratropical cyclone over Virginia on September 28, before moving back into the ocean. Frances, a Category 2 storm, moved into the Gulf of Mexico.

The charts offer some clues as to why the storms turned in different directions. For Jeanne, Mercury conjunction the Midheaven indicated winds blowing from the south to the north, and the square to Pluto indicated an easterly direction; the storm traveled to the northeast. The ingress chart for Frances also suggests a post-landfall direction. Lunar phase Saturn at 23 Cancer on the western side of the chart was square the ingress meridian, about 2 degrees from exact. Ingress charts for areas to the northwest had lower degrees of Aries/Libra on the meridian, and Frances followed the Saturn "path" to Tampa, which had 23 Aries/Libra on the meridian. You can use this technique to narrow the forecast cone posted on the Web by the National Hurricane Center. And because warm water is a hurricane's fuel, it's also helpful to check the aspects of warm planets (Mars, Venus, Neptune) to offshore ocean locations along with Sea Surface Temperature (SST) maps, also on the National Hurricane Center website.

GALVESTON, TEXAS, SEPTEMBER 8, 1900

One of the most devastating hurricanes to hit the United States powered into Galveston, Texas, on September 8, 1900. Deaths were estimated at 10,000 to 12,000, with 6,000 in the city itself. As the eye of this Category 4 hurricane approached, winds reached 110 mph and a five-foot water surge charged onto land. Because buildings had been erected on sand, many toppled as the water and rain, driven by high winds, pounded them. Every building within three blocks of the Gulf Shore was destroyed, with a total of 3,600 structures leveled in the storm's path. Unfortunately, many of the people killed had taken refuge in the buildings when the storm hit.

Rain began early on the morning of September 8 at 7:45 am, accompanied by a plunging barometer that fell to a low of 945 mb. By noon it was raining heavily, the wind was blowing at 40 mph, and four feet of water stood in the streets. That afternoon, winds increased and rain fell in torrents. When the leading edge of the eye crossed the shoreline that evening at about 7:00 pm, the initial water surge of four feet was followed by another of twelve feet. It was only the beginning. The storm continued all night.

This example illustrates how you can take a closer look at hurricane timing, including the beginning of rain, the intensification of rain and wind, landfall, and the arrival of the eye.

The June 21, 1900, summer ingress (inner wheel of Hurricanes 8) shows hurricane potential with Mercury sesquisquare Uranus. However, with only Uranus aspecting the Ascendant (semisextile), it would take a lunar phase to activate the potential.

At the ingress, Mars was semisextile Neptune, but only Neptune aspected the horizon (sesquisquare the Ascendant). Saturn was strong, aspecting the horizon and meridian, as well as forming an opposition with the Sun (easterly winds, slowly falling barometer, and intense rain with thunder), a contraparallel with Mercury (easterly winds, falling barometer, and rain) and Neptune (counterclockwise low-pressure system, heavy rain, and humid), and a parallel with Uranus (high wind and heavy rain). The lunar eclipse of June 13, 1900, at 21 Sagittarius was sextile the ingress IC, and a significator of unusual weather was present with Uranus contraparallel Neptune.

Venus, which had turned retrograde four days before the ingress, was conjunct Mercury in the ninth house in the water-producing sign Cancer. It was conjunct Mercury, and both planets were contraparallel Saturn and sesquisquare and contraparallel Uranus, increasing the potential for damp, cold rain, as did their contraparallel with Jupiter, sextile to Mars, and semisextile to Neptune.

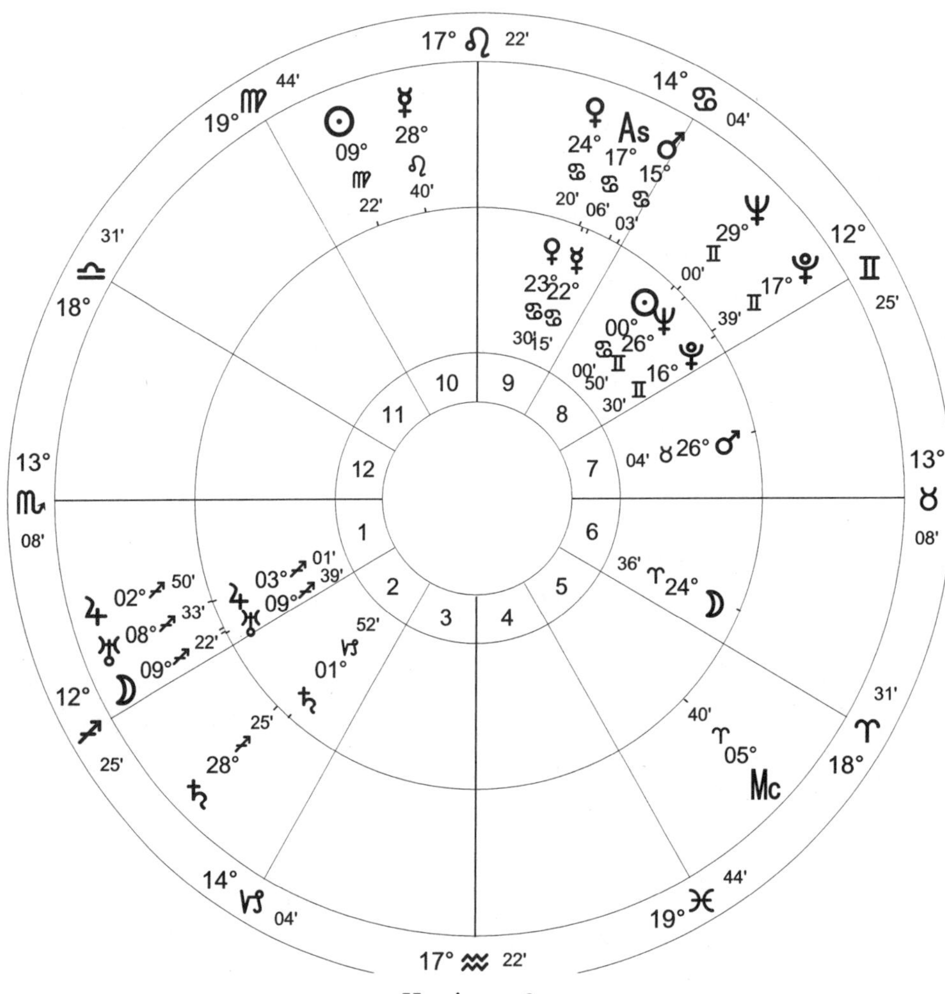

Hurricanes 8

INNER WHEEL=*Summer Ingress / Natal Chart / June 21, 1900 NS, Thu / 3:40 pm CST +6:00*
Galveston, TX / 29°N18' 094°W48' / Geocentric / Tropical / Placidus / Mean Node

OUTER WHEEL=*Lunar Phase / Natal Chart / September 2, 1900 NS, Sun / 1:56 am CST +6:00*
Galveston, TX / 29°N18' 094°W48' / Geocentric / Tropical / Placidus / Mean Node

At the lunar phase of September 2, 1900 (outer wheel of Hurricanes 8), Saturn formed an opposition to Neptune, adding fuel to the contraparallel between the two planets, and a parallel with Uranus. This time, Mercury was trine Saturn (easterly counterclockwise winds and falling barometer), and Saturn activated the ingress Ascendant by semisquare as it turned stationary direct on September 4: a major low-pressure system occurred that week. The solar eclipse of May 28, 1900, at 6 Gemini was trine/sextile the lunar phase meridian.

It would have been easy to overlook the hurricane potential in the lunar phase chart because Mercury formed no aspect to Uranus, as it did in the ingress chart. But by September 8, Mercury had moved forward to square that planet, which was trine/sextile the meridian and parallel/scontraparallel the horizon.

Unlike in the ingress chart, lunar phase Mars formed five aspects that contributed to the hurricane: sesquisquare Jupiter (thunderstorm), contraparallel Saturn (windy storm), parallel Neptune (rapidly falling barometer, hurricane, torrential rain), semisquare Mercury (thunderstorms), and semisextile Pluto (wind). Mars was conjunct and parallel the lunar phase Ascendant and semisquare/sesquisquare the ingress meridian. Thunderstorms and wind were also indicated by the Mercury-Pluto parallel in the lunar phase chart and by the transiting Sun square the ingress Pluto.

Ingress retrograde Venus, after turning direct, had returned to its ingress place at the lunar phase, indicating rain in its sesquisquare to ingress Uranus, square to ingress Mars, and conjunction with ingress Mercury. Ingress Uranus sesquisquare lunar phase Venus added to the rain potential, as did transiting Venus semisquare/sesquisquare the horizon. Lunar phase Uranus was sesquisquare ingress Mercury, and the lunar phase Sun was square ingress Uranus within 18 minutes of exactitude.

The hurricane's arrival in Galveston was imminent on the evening of September 7, 1900, when transiting Mercury at 9 Virgo squared ingress Uranus and formed these aspects to lunar phase chart planets: semisquare Venus, square the Moon, and conjunct the Sun. The transiting Moon in Pisces trined the ingress Sun, setting off its low-pressure opposition to Saturn and rain-making conjunction with Neptune. A series of planets in the lunar phase chart was aspected by the transiting Moon in Aquarius and Pisces: contraparallel Sun, sextile Saturn, opposition Mercury, trine Neptune, and sesquisquare Mars. Later that evening, the transiting Sun at 15 Virgo was sextile lunar phase Mars.

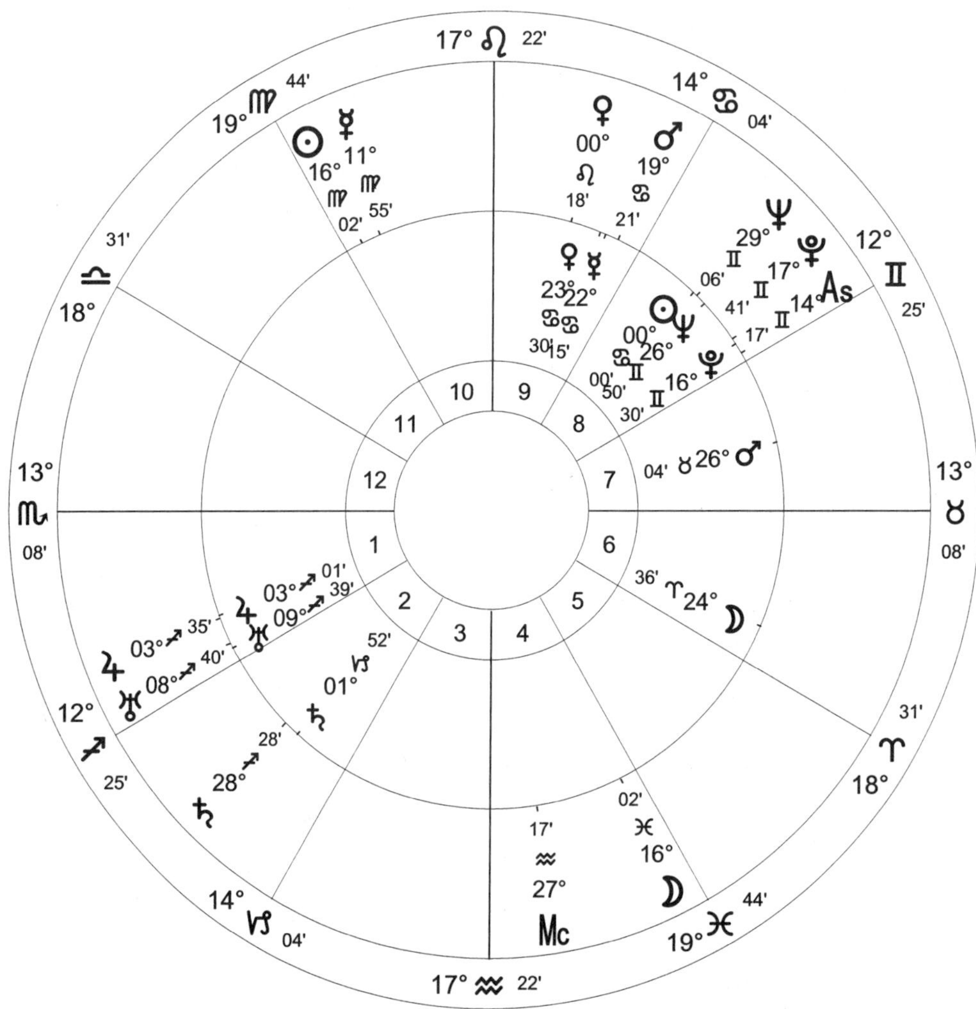

Hurricanes 9

Inner Wheel=*Summer Ingress / Natal Chart / June 21, 1900 NS, Thu / 3:40 pm CST +6:00*
Galveston, TX / 29°N18' 094°W48' / Geocentric / Tropical / Placidus / Mean Node

Outer Wheel=*Lunar Phase / Natal Chart / September 8, 1900 NS, Sat / 11:06 pm CST +6:00*
Galveston, TX / 29°N18' 094°W48' / Geocentric / Tropical / Placidus / Mean Node

Soon after midnight, the transiting Moon was sextile ingress Saturn, further intensifying the low pressure, and sesquisquare the lunar phase Ascendant. Around noon on September 8, 1900, the Moon sesquisquare the ingress Mercury-Venus conjunction reflected heavy rain, as did the transiting Moon's sesquisquare to lunar phase Venus. The transiting Moon made aspects to planets in the lunar phase chart as well: parallel the IC, square Uranus, and sesquisquare Venus before the arrival of the full Moon at 3:40 pm on September 8.

In the September 8 lunar phase chart (outer wheel of Hurricanes 9), Pluto was conjunct the Ascendant (high wind), being squared by the Moon in the tenth house and the Sun in the fourth. Saturn in the seventh house was in a nearly exact opposition to Neptune in the first house, with both of them aspecting the meridian. Mercury was square Uranus. The full fury of the hurricane was thus unleashed in the early evening as the transiting Moon moved into an exact square to Pluto and the transiting Sun to a semisquare with Venus.

The wind and rain further escalated late that afternoon with the Moon's square to ingress Uranus and Venus's semisextile to the ingress Sun. The eye arrived about 7:00 pm as the Moon was trine/sextile the horizon, and the wind continued to howl into the night as the Moon squared ingress Pluto and formed an inconjunct/semisextile with the meridian. The ingress Moon and Pluto were parallel the lunar phase IC, and the ingress Sun, Mercury, and Neptune were parallel the lunar phase Ascendant.

HURRICANE FREDERIC, SEPTEMBER 12–13, 1979

Hurricane Frederic crossed the coastline near the entrance to Mobile Bay, Alabama, on September 12, 1979, destroying or severely damaging most beachfront homes. Maximum wind speeds were 144 mph, with flood waters rising to between eleven and fourteen feet. Ninety to ninety-five percent of the power was out in Mobile County, tornadoes were reported, 500,000 people were evacuated from a 100-mile area from Gulfport, Mississippi, to Pensacola, Florida, and schools were closed for two weeks. Nine people died, and damages reached $1.5 billion. Frederic then turned northward, finally exiting the North American continent in northern Maine. As it traveled between Alabama and Maine, it caused $2.3 billion in property damage and thirteen people lost their lives.

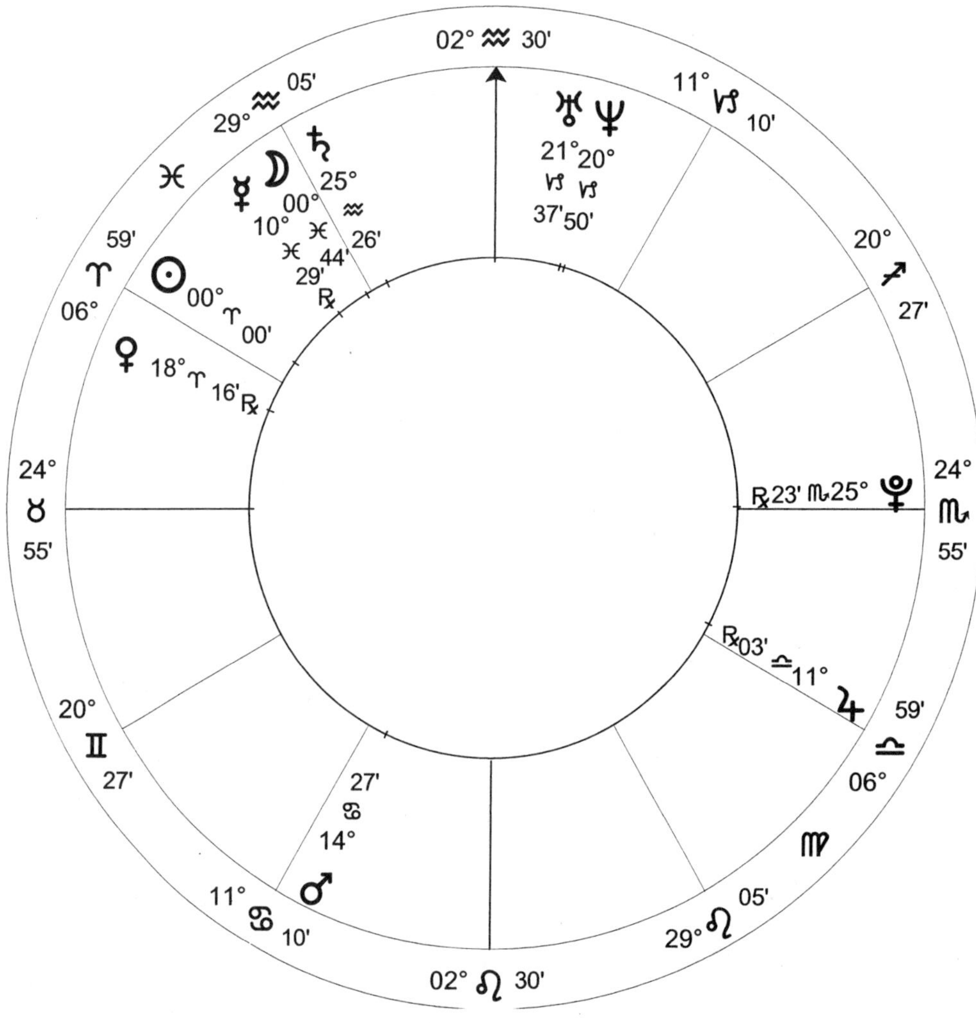

Hurricanes 10

Inner Wheel=*Summer Ingress / Natal Chart / June 21, 1979, Thu / 6:57 pm CDT +5:00*
Mobile, AL / 30°N41'39" 088°W02'35" / Geocentric / Tropical / Placidus / Mean Node

Outer Wheel=*Lunar Phase / Natal Chart / September 6, 1979, Thu / 5:59 am CDT +5:00*
Mobile, AL / 30°N41'39" 088°W02'35" / Geocentric / Tropical / Placidus / Mean Node

The June 21, 1979, summer ingress chart (inner wheel of Hurricanes 10) shows the potential for heavy rains and flooding with Neptune conjunct the Ascendant and two other outer planets, Uranus and Pluto, also aspecting the ascending degree. These three planets, plus one more, were the key factors in a powerful hurricane arriving on the shores of Mobile.

The ingress chart indicated the potential for a low-pressure system with counter-clockwise winds because Mercury was semisquare Saturn and Venus, with all planets in high declination; Saturn parallel Pluto suggested a storm with high winds. Thunder-storms manifested when the Mars parallel Jupiter and Venus sextile Jupiter aspects were activated. Venus sextile Jupiter generally indicates fair weather, but when either planet is aspected by Mars, as is the case here, the combination has thunderstorm potential. In addition to Neptune's placement, humid, wet weather was sure to be the norm with Sun parallel Mercury and sesquisquare Uranus; Mercury semisquare Saturn; Venus square Saturn, contraparallel Neptune, and trine Pluto; and Neptune sextile Pluto. The lunar eclipse of September 6, 1979, at 13 Pisces was square the Ascendant, and the solar eclipse of August 22, 1979, at 29 Leo formed an inconjunct with the IC.

Nevertheless, the planetary configurations in the ingress chart gave little clue that Mobile would find itself in the middle of Hurricane Frederic in September. Absent were the classic Mercury, Mars, Uranus, and Neptune hurricane aspects. Only Mercury semis-quare Saturn showed the potential for counterclockwise air flow, but hardly enough to forecast a hurricane. This emphasizes the importance of studying the lunar phase charts and their aspects to the ingress chart. Nevertheless, because Mobile is hurricane-prone and because Neptune is conjunct the ingress Ascendant, it would be wise to be alert for lunar phases and hard aspect transits to Mercury-Saturn.

But before looking at the lunar phase chart for September 6, 1979, let's explore Fred-eric's roots. The summer ingress chart for the Cape Verde Islands area off the coast of Africa (Hurricanes 11) showed major potential for the formation of low-pressure storms with counterclockwise winds. When this ingress chart is rotated to reflect the longitude, latitude, and time zone of the Cape Verde Islands, 0 Pisces/Virgo is on the horizon and 7 Sagittarius/Gemini is on the meridian. Venus was parallel the IC (precip-itation), and Uranus (high pressure) was semisextile Neptune in the tenth and contra-parallel the IC (precipitation). Mars was square the horizon, while Saturn (low pressure) was in a nearly exact square to the meridian. Conditions were ripe for development of many strong low-pressure systems with counterclockwise winds, some of which would advance from tropical depression to full hurricane. The solar eclipse of August 22, 1979, at 29 Leo formed an inconjunct/semisextile with the horizon, and the lunar eclipse of September 6, 1979, at 13 Pisces was square the meridian.

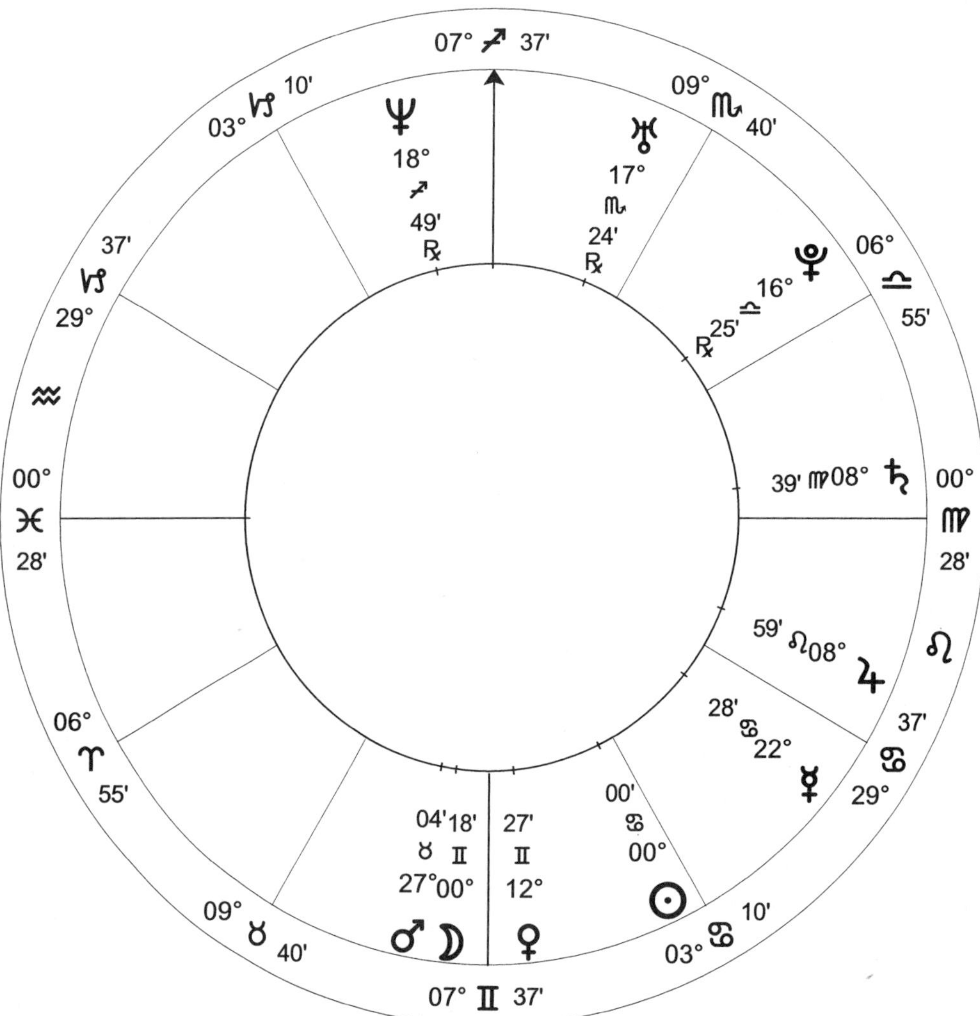

Hurricanes 11: Summer Ingress / Natal Chart / June 21, 1979, Thu / 9:57 pm AZ +2:00
Cape Verde Islands / 16°N00' 023°W00' / Geocentric / Tropical / Placidus / Mean Node

A look at the transiting planets in the weeks before Hurricane Frederic arrived in Mobile shows a number of significant aspects. The Sun, Venus, and Mercury all transited 7 Virgo, setting off the Cape Verde ingress Saturn square to the meridian and, subsequently, ingress Venus and Neptune. Transiting Saturn, advancing in degree, was narrowing the orb in its approaching square to ingress Neptune, an aspect that would be exact when Frederic hit Mobile. At the end of August, transiting Mercury conjunction Jupiter in Leo was semisquare the ingress Saturn and meridian, providing the heat necessary to begin fueling what would become Hurricane Frederic. The necessary warm water is reflected in transiting Mars inconjunct Neptune.

As the storm developed into Hurricane Frederic on its trek across the Atlantic Ocean, the lunar phase chart of September 6, 1979 (lunar eclipse), for Mobile showed that the city would be its target (outer wheel of Hurricanes 10). Mercury conjunction the Ascendant and ingress Saturn and square the meridian promised a low-pressure system with high winds. Neptune parallel the IC and square Venus and Saturn confirmed the potential for a major low-pressure system with heavy precipitation and counterclockwise winds. This was aggravated by the Sun conjunction and parallel Venus and Saturn, the former also adding to the heat.

Mars in high declination had a role as well, with its semisquare to Mercury (thunderstorm, whipping winds), trine to Uranus (overcast with wind gusts), inconjunct to Neptune in high declination (humidity, low pressure), and square to Pluto (high wind). Uranus and Neptune were still aspecting each other, indicating unusual weather. Mars was semisquare/sesquisquare the meridian and semisquare/sesquisquare the horizon, and Pluto was semisquare/sesquisquare both the horizon and the meridian.

However, the key factors in the immense storm were the four outer planets—Saturn, Uranus, Neptune, and Pluto—all in aspect to each other at the same degree in different signs. By September 12, Saturn had moved forward to 17 Virgo to complete the quartet.

Connections between the lunar phase and ingress charts also set things in motion. In addition to lunar phase Mercury forming a conjunction with ingress Saturn, Mercury was semisquare Mercury, lunar phase Sun was square ingress Venus, and lunar phase Venus was square ingress Neptune. Uranus had returned to within sixteen minutes of its position in the ingress chart, and lunar phase Mars formed aspects with four ingress chart planets: conjunction Mercury, trine Uranus, inconjunct Neptune, and square

Pluto. Lunar phase Mars also was contraparallel/parallel the ingress horizon. Lunar phase Jupiter was square ingress Mars, forming the primary thunderstorm-breeding configuration. Four ingress planets—Sun, Mercury, Venus, and Neptune—aspected the lunar phase meridian by parallel or contraparallel.

On the day Frederic arrived, the transiting planets formed ominous aspects with each other and with the lunar phase and ingress charts. Transiting aspects, all separating, were Sun at 19 Virgo conjunct Saturn at 17 Virgo (low pressure, precipitation), Sun square Neptune at 17 Sagittarius (hot, humid, squall), Mercury at 19 Virgo conjunct Saturn (falling barometer, precipitation, leaden skies, easterly and counterclockwise winds), Mercury sextile Uranus at 18 Scorpio (high pressure, erratic westerly wind gusts, descending cold air), Mercury square Neptune (low pressure, leaden skies, tornadoes, squall), Mercury semisextile Pluto at 18 Libra (westerly winds), and Sun semisextile Pluto (heat, thunderstorms). Note that each of these aspects, although separating, involved one of the four outer planets.

Other transiting aspects that day were Sun parallel Venus (hot, humid, rain), Mercury parallel Saturn, and Mars contraparallel Neptune (both in high declination), to create low pressure, rapidly falling barometer, rising very moist air, counterclockwise winds—a hurricane.

The transiting planets contacted the ingress chart, forming more aspects to add to Frederic's intensity: Sun square Neptune (hot, humid, squall); Mercury square Neptune (low pressure, humid, tornadoes); Jupiter square Mars (thunderstorms); and Mars conjunct and parallel Mercury (westerly whipping winds, thunderstorms), semisquare Saturn (destructive windy storm), parallel Venus (thunderstorms), and contraparallel Neptune.

These were the aspects the transiting planets made to the lunar phase chart: Saturn exactly square Neptune, exactly parallel Venus, sextile Uranus, and semisextile Pluto (cloudy, windy); Mars semisquare Mercury; and Mercury parallel Saturn and contraparallel the Moon (windy), parallel the Sun (showers) and Venus (showers), and sextile Mars (westerly winds). Note again how the transiting planets aspected the four outer planets.

With so many outer planet contacts (and others that were separating), the question arises as to why Frederic unleashed its fury on Mobile on September 12 instead of a few days earlier. The transiting Moon on September 12 was in Gemini and triggered the outer planets, all at the same degree and in aspect with one another, with the Saturn-

Neptune square as the prime force. This illustrates the pivotal role the transiting Moon can play in timing weather. It was the Moon, the swiftest-moving planet, that indicated Frederic's arrival on the shores of Mobile on September 12, 1979.

Saffir-Simpson Hurricane Wind Scale

The Saffir-Simpson hurricane wind scale assigns a rating between 1 and 5 to each hurricane. Each of the five categories is defined by the sustained (not gusts) wind velocity and level of wind damage caused by the hurricane.

Category 1

Sustained Winds: 74–95 mph

Damage: Very dangerous winds will produce some damage. Well-constructed frame homes could have damage to roof, shingles, vinyl siding, and gutters. Large branches of trees will snap, and shallowly rooted trees may be toppled. Extensive damage to power lines and poles likely will result in power outages that could last a few to several days.

Category 2

Sustained Winds: 96–110 mph

Damage: Extremely dangerous winds will cause extensive damage. Well-constructed frame homes could sustain major roof and siding damage. Many shallowly rooted trees will be snapped or uprooted and block numerous roads. Near-total power loss is expected, with outages that could last from several days to weeks.

Category 3

Sustained Winds: 111–129 mph

Damage: Devastating damage will occur. Well-built framed homes may incur major damage or removal of roof decking and gable ends. Many trees will be snapped or uprooted, blocking numerous roads. Electricity and water will be unavailable for several days to weeks after the storm passes.

Category 4

Sustained Winds: 130–156 mph

Damage: Catastrophic damage will occur. Well-built framed homes can sustain severe damage, with loss of most of the roof structure and/or some exterior walls. Most trees will be snapped or uprooted and power poles downed. Fallen trees and power poles will isolate residential areas. Power outages will last weeks to possibly months. Most of the area will be uninhabitable for weeks or months.

Category 5

Sustained Winds: 157 mph or higher

Damage: Catastrophic damage will occur. A high percentage of framed homes will be destroyed, with total roof failure and wall collapse. Fallen trees and power poles will isolate residential areas. Power outages will last for weeks to possibly months.

CHAPTER 9

Snow, Sleet, and Freezing Rain

Snow, sleet, and freezing rain are all common forms of precipitation in northern latitudes. Winter sports enthusiasts look forward to the first snowfall, while commuters everywhere dread seeing any of the three falling from the sky.

Snow formation requires two ingredients: a sufficient supply of cold air and moisture. When the cold air descends far south, usually as a result of a jet stream that dips into more southerly latitudes, people living in areas normally receiving only rain can find themselves in the midst of all three forms of winter precipitation. Conversely, when cold air and the jet stream remain far to the north, areas where snow is the norm might experience only rain or sleet.

Major snowstorms and blizzards usually are the result of an extratropical cyclone, which has a distinctive "comma" shape and looks similar to a hurricane when viewed on satellite imagery. An extratropical cyclone can produce rain on the southern and eastern sides (warm sector) and snow on the northern and western sides (cold sector), with the heaviest precipitation on the northern side, northwest of the low pressure. The cyclone has a counterclockwise movement, like a hurricane, and is usually steered from west to east by strong westerly winds. Barometric pressure is lowest at the center of the cyclone, and tight pressure gradients result in high winds.

Snow storms form in varying patterns and at varying degrees of intensity across the North American continent. Some begin in the Pacific Ocean, cross the West Coast, and continue until they reach the Rocky Mountains. If the storm crosses the Rockies and enters the Plains, it begins to gain strength, drawing moisture from the Gulf that mixes with cold, dry air descending southeastward from Canada. These contrasting air masses supply the needed components.

As snow storms make their way across the Plains, some veer northward to Canada, crossing the Great Lakes. Others continue their path eastward and seem to die as they encounter the Appalachian Mountains. However, sometimes the cold air and counter-clockwise winds in the upper atmosphere move into the Northeast or across the Appalachians to create a "secondary" storm, resulting in some of the heaviest snowfalls in history along the Atlantic Coast. The same area also experiences snow when warm, moist air rises along the Atlantic Coast and meets cold air descending from the north; this can result in a nor'easter.

Rain can begin as snow and melt when it hits warm air closer to Earth. Freezing rain results when snow melts on its descent through warmer air, then cools in cold air and forms ice as it hits the colder ground. When snow melts and refreezes in colder air before hitting the earth, sleet—actually small ice pellets—results. (Often an area will receive both sleet and freezing rain, the two alternating throughout the storm with slight variations in atmospheric conditions.) Snow remains snow when it falls through only cold air.

The first component of a snow storm—moisture—is provided by Venus, Neptune, or planets in water signs. Add to that the ingredient of cold air, which comes from Uranus, Saturn, Mercury, or planets in the cold signs of Capricorn and Aquarius. The attendant clouds are indicated by, for example, Venus in aspect with Saturn.

Another required element of a snow storm is crystallization of water vapor. Without it, the condensing water vapor does not form into an ice crystal (snowflake). Maximum crystallization occurs when Venus (moisture) is in Aquarius (the coldest and most crystallizing sign) or when Saturn or Uranus forms a sextile with another planet. When Venus in Aquarius is sextile Saturn or Uranus, expect snow, especially if one or more of the planets forms a hard aspect with the meridian at a specific location.

The Sun in negative or neutral aspect to Mercury, Venus, Saturn, or Uranus during winter months indicates snow. However, it is the added influence of Mercury, planet of wind, that brings the drifts associated with many snow storms. If retrograde Mercury forms a hard or neutral aspect with the Sun and is appropriately positioned in the chart, a blizzard is likely. Retrograde Venus is also associated with blizzards.

A snow storm becomes a blizzard when temperatures remain below 25 F and wind velocity is 35 mph or higher. Thus a blizzard requires cold air and high-velocity winds, as well as moisture. Mercury produces high winds when in aspect with Mars; if it negatively aspects Pluto, the wind is intensified. Major snow storms, and especially blizzards, are often indicated by the presence of Pluto (high wind) conjunct or square the meridian.

A major snow storm or blizzard requires a constant and extended flow of moisture. When in hard aspect, Venus and Neptune can indicate heavy precipitation, as does the Sun with Venus or Mercury. Planets in hard aspect to Saturn also increase the potential. Storms strongly influenced by Saturn are slow to develop and dissipate, and Saturn provides the low pressure (as does Neptune) component of a storm. Saturn is usually active in an extratropical cyclone.

BLIZZARD, JANUARY 25–27, 1978

This blizzard buried locations in the Ohio Valley and Great Lakes under heavy snow. Some of the totals: Detroit, eight inches; South Bend, Indiana, three feet; Chicago, thirteen inches; Lansing, Michigan, nineteen inches; and Indianapolis, fifteen inches. About twenty people died as a result of the storm, and many were hospitalized due to exposure. About 100,000 vehicles were abandoned in Michigan alone, and airports were closed. Temperatures in Indianapolis dropped to 0 F, and drifts were ten to twenty feet high. The charts shown here are for Indianapolis and are representative of those across the region.

In the December 21, 1977, winter ingress chart (inner wheel of Snow 1), the Sun was conjunct retrograde Mercury, a blizzard significator. The Sun also activated Uranus by semisquare, an indicator of wind and snow. Still more wind was indicated by retrograde Mercury sesquisquare and contraparallel Mars, opposition Jupiter, trine Saturn, semisquare Uranus, and parallel Neptune (low pressure). To this was added the storm indicators of Mars square Uranus (wind) and contraparallel Neptune (heavy precipitation, low pressure). Venus was sesquisquare Mars and contraparallel Jupiter (precipitation).

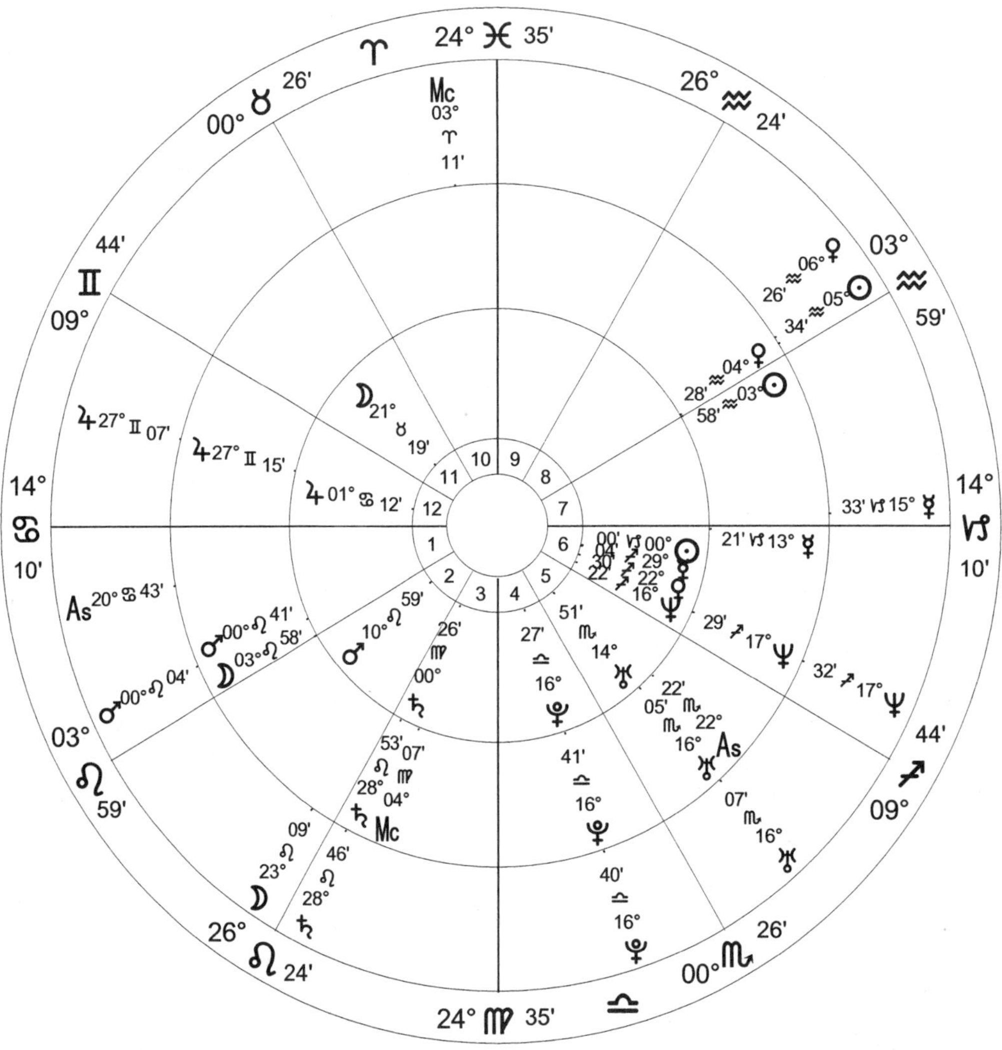

Snow 1

INNER CHART=*Winter Ingress / Natal Chart / December 21, 1977, Wed / 6:23:09 pm EST +5:00*
Indianapolis, IN / 39°N46'06" 086°W09'29" / Geocentric / Tropical / Placidus / Mean Node

MIDDLE WHEEL=*Lunar Phase / Natal Chart / January 24, 1978, Tue / 2:55:30 am EST +5:00*
Indianapolis, IN / 39°N46'06" 086°W09'29" / Geocentric / Tropical / Placidus / Mean Node

OUTER WHEEL=*Snow Begins / Natal Chart / January 25, 1978, Wed / 4:37 pm EST +5:00*
Indianapolis, IN / 39°N46'06" 086°W09'29" / Geocentric / Tropical / Placidus / Mean Node

This combination signals blizzard potential because it incorporates all the necessary planetary factors and aspects: Mercury (wind), Neptune (southerly, moist air), Uranus (cold, descending air, high pressure), Venus and Neptune (precipitation, moist southerly air), Saturn (cloudy, cold, low pressure), and Pluto (wind). Venus was square the meridian, Mars was sesquisquare/semisquare the meridian and semisextile the Ascendant, Neptune was inconjunct the Ascendant, and Saturn was semisquare the Ascendant.

The January 24, 1978, lunar phase chart (middle wheel of Snow 1) also had strong indicators for a blizzard. Uranus (cold) was conjunct the Ascendant, sextile Mercury (wind), semisextile Neptune (precipitation) and Pluto (wind), and sesquisquare Saturn (low pressure). Even more important and indicative of a blizzard was retrograde Mars opposition the Sun-Venus conjunction in Aquarius. Mars was sesquisquare Neptune (storm, low pressure), and the Sun-Venus conjunction was semisquare Neptune.

Lunar phase Mercury was conjunct the ingress horizon, the lunar phase Midheaven was conjunct ingress and lunar phase Saturn, and the lunar phase Ascendant was trine the ingress Midheaven. All of these aspects focused the blizzard indicators on the longitude and latitude of Indianapolis.

The transiting Moon was a good timer for this storm. Snow began falling as the Moon formed a trine to ingress Venus and an inconjunct/semisextile to the meridian. The next morning (January 26), the barometric pressure fell to 958 mb (average pressure ranges from about 980 to 1003, depending on location and low- and high-pressure systems in effect). This dramatic pressure drop, something that is usually only associated with a strong hurricane, occurred as the transiting Moon formed a conjunction with transiting/lunar phase/ingress Saturn.

THE BLIZZARD OF 1888

Possibly the worst nor'easter in recorded history was the Blizzard of 1888, which hit the northeastern United States on March 12, 1888. It extended from Maine to Maryland and from Buffalo, New York, to Pittsburgh, Pennsylvania, dropping as much as four feet of snow in some areas. Drifts towered to thirty and forty feet, with 50 mph winds; the temperature dropped to 10 F. In New York City, there were 200 deaths, and houses were buried in drifts of fifteen to twenty feet. Throughout the affected area, 400 people died, including 100 seamen who were on 200 vessels that sank.

Snow began falling just after midnight on March 12, 1888, after several days of unseasonably mild temperatures, and tapered to a light snowfall in the early morning hours of March 13.

The spring-like weather preceding the blizzard was the result of warm, moist air moving northward along the Atlantic Coast. When this warm air encounters cold, dry Arctic air moving south from Canada, a low-pressure system develops, and its counterclockwise air flow moves the warm, moist air over land. The greater the temperature difference, the deeper the low, the higher the winds, and the more intense the storm. Nor'easters thrive on cold air.

The December 21, 1887, winter ingress chart for New York City (inner wheel of Snow 2) set the scene for what was to come the following March. Snow and colder temperatures were indicated by the Sun semisquare Venus in wet Scorpio. Jupiter in Scorpio opposition and contraparallel Neptune indicated above-normal precipitation and low pressure. Pluto conjunct the Midheaven and square the horizon, spurred on by the sesquisquare from cold Uranus, indicated high winds and storms, while Uranus sesquisquare Neptune showed potential for heavy precipitation and unusual weather.

The Blizzard of 1888 is especially interesting because it occurred during two lunar phases, the end of one and the start of another. The first lunar phase (outer wheel of Snow 2) set the blizzard in motion, and the second (outer wheel of Snow 3) provided the fuel that secured the storm's place in the record books. The lunar eclipse of January 28, 1888, at 8 Leo was semisextile the ingress Ascendant, trine the ingress IC, square the March 4, 1888, lunar phase Ascendant, and semisextile the March 12, 1888, lunar phase IC. The March 4, 1888, lunar phase IC was conjunct the solar eclipse of February 11, 1888, at 22 Aquarius.

The March 4, 1888, lunar phase chart is an ominous one for a potential blizzard category because of retrograde Mercury conjunct the Sun in wet Pisces and sesquisquare Saturn. The classic destructive windy storm breeders, Mars (warm air) and Saturn (cold air), were in an exact square, and both were retrograde.

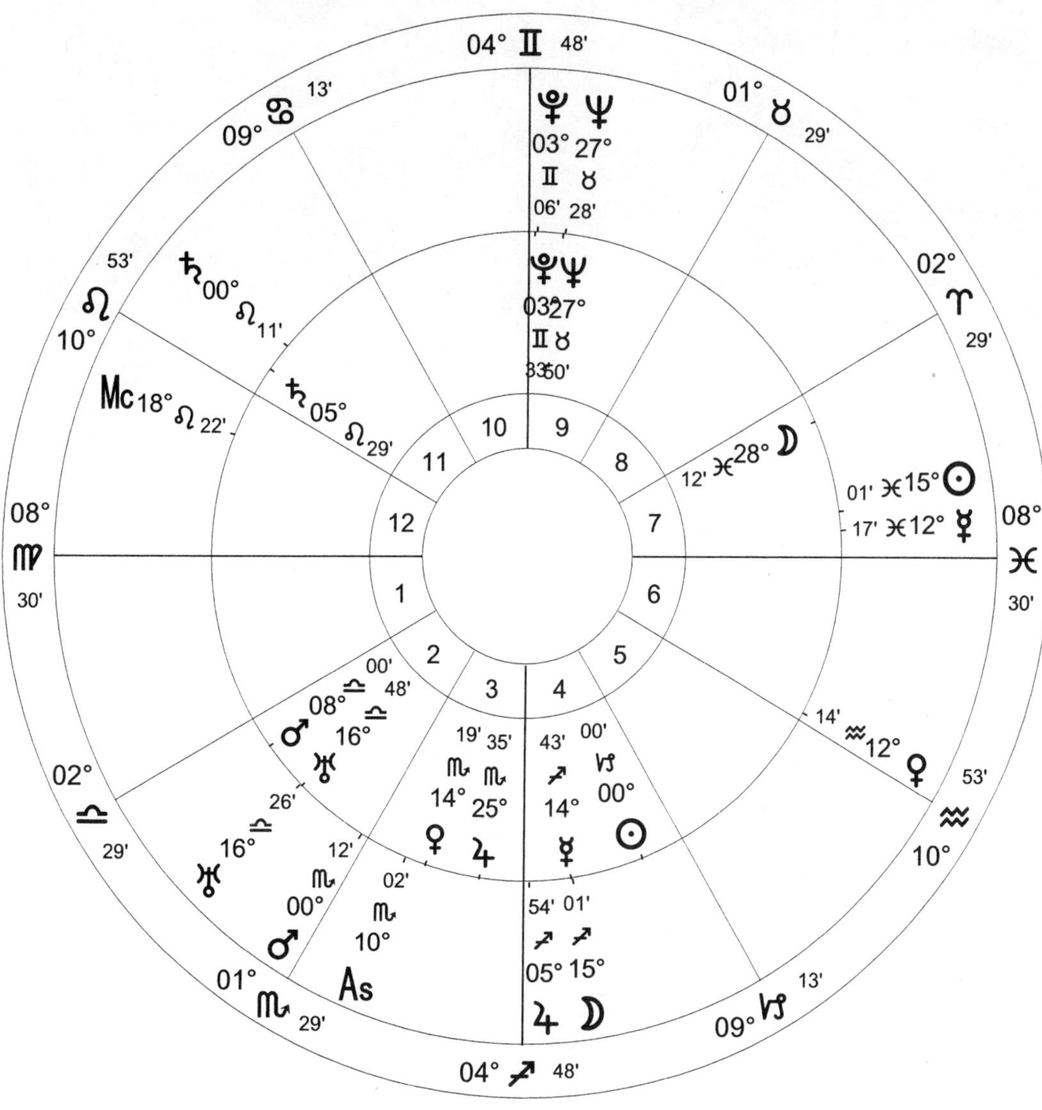

Snow 2

INNER WHEEL=*Winter Ingress / Natal Chart / December 21, 1887 NS, Wed / 10:06 pm EST +5:00*
New York, NY / 40°N42'51" 074°W00'23" / Geocentric / Tropical / Placidus / Mean Node

OUTER WHEEL=*Lunar Phase / Natal Chart / March 4, 1888 NS, Sun / 10:26 pm EST +5:00*
New York, NY / 40°N42'51" 074°W00'23" / Geocentric / Tropical / Placidus / Mean Node

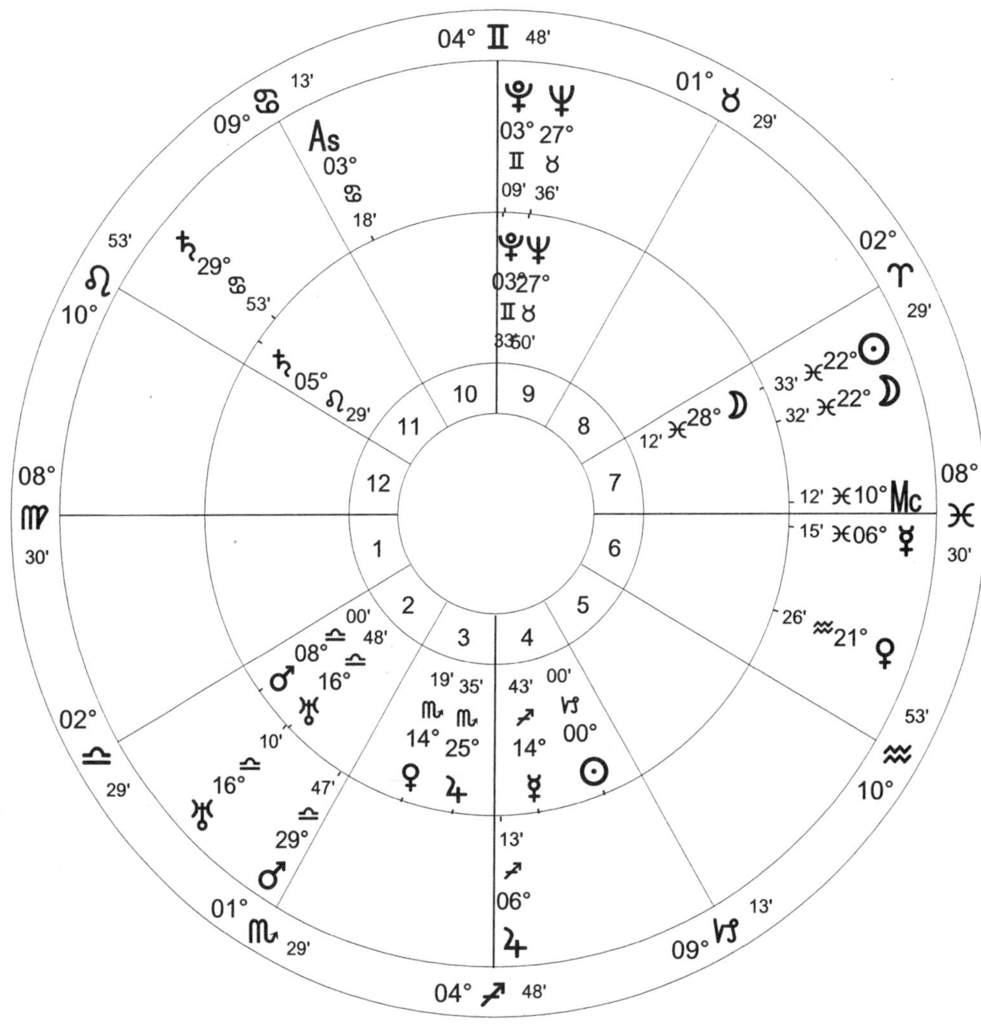

Snow 3

INNER WHEEL=*Winter Ingress / Natal Chart / December 21, 1887 NS, Wed / 10:06 pm EST +5:00*
New York, NY / 40°N42'51" 074°W00'23" / Geocentric / Tropical / Placidus / Mean Node

OUTER WHEEL=*Lunar Phase / Natal Chart / March 12, 1888 NS, Mon / 11:20 am EST +5:00*
New York, NY / 40°N42'51" 074°W00'23" / Geocentric / Tropical / Placidus / Mean Node

The lunar phase Sun sesquisquare Saturn, Venus contraparallel Neptune and square the horizon, and Jupiter contraparallel Saturn indicated heavy precipitation. The Saturn aspects, including the sesquisquare from Mercury, contributed to the cold temperatures indicated by the Sun parallel Uranus, as did Sun conjunct Mercury, Venus in Aquarius square the horizon, and Uranus sextile/trine the meridian. Through an opposition, Jupiter expanded Pluto's tendency for high wind, and also represents the contrast between the warm, southerly and the cold, northerly air flows that are characteristic of a nor'easter.

Both Venus and Mercury in the March 4, 1888, lunar phase chart were square their ingress positions, identifying this week as one in which the ingress aspects would be activated. The March 4 lunar phase chart had Sun square ingress Mercury, providing another trigger for retrograde Mercury, which served as the focal point to kick off the blizzard. Lunar phase Mars was semisquare ingress Mercury (high winds), a planet that also was sesquisquare ingress Saturn (overcast, precipitation, counterclockwise winds).

When March 12 arrived, the Moon in Pisces activated ingress Mercury and Venus and the Uranus-Neptune and Uranus-Pluto sesquisquares in both charts. The Moon continued its transit through Pisces throughout the day and into the early morning hours of March 13, aspecting transiting Saturn, Neptune, Mars, and Uranus. The wind howled through the night as transiting Mars was semisquare ingress Mercury.

Planetary positions at the March 12, 1888, lunar phase added fuel to the raging blizzard with Mars square Saturn (windy storm), and Mercury forming a T-square with Jupiter and Pluto (high wind and the contrasting air flow). Lunar phase Venus was sesquisquare ingress Mars and square ingress Jupiter, indicating a source of warm, moist air. Lunar phase Venus was also sesquisquare ingress Mars, further indicating a southerly flow of air. Cold air is indicated by ingress Uranus parallel the lunar phase IC, which was also parallel ingress Pluto.

The ingress and lunar phase (March 4) planets occupying the lower meridian pinpointed the Northeast as the site of the blizzard, as did the ingress Ascendant conjunct the March 12 lunar phase IC; the ingress Ascendant was opposed by the March 4 lunar phase Mercury, and there was a Mercury-Midheaven conjunction in the March 12 lunar phase chart. Jupiter in both lunar phase charts aspected the ingress meridian and horizon, and Saturn (both lunar phases) was contraparallel the ingress IC, concentrating the low pressure over the area. The ingress Sun was contraparallel the March 12 lunar phase

Ascendant, lunar phase (both phases) Mars was contraparallel the ingress Ascendant, and lunar phase (both phases) Pluto was parallel the ingress Ascendant. The March 12 lunar phase IC was parallel the ingress Ascendant.

Chicago, Illinois, January 26–27, 1967

Chicagoans awakened Thursday morning, January 26, 1967, to a small accumulation of snow that had begun falling at 5:02 am. However, by mid-morning, four to five inches covered the ground, and by 6:00 pm residents were trying to navigate through a foot of snow. When the snow stopped falling the next morning at 10:10 am, twenty-three inches had accumulated. Northeast winds that gusted to 53 mph created drifts of four to six feet. Temperatures ranged from 26 to 29 F. Business losses were $150 million, and snow removal costs for the winter soared to $5 million. Sixty people lost their lives in the blizzard.

The winter ingress chart (inner wheel of Snow 4) combined several key components of destructive weather conditions. The Sun parallel Venus (snow and colder) received added fuel from Venus square Mars to transform the flakes into a storm. Other indicators of snow were Venus semisquare Neptune and Saturn trine Neptune. The solar eclipse of November 12, 1966, at 20 Scorpio was semisextile the Ascendant.

But what indicated the potential for a major, crippling storm was Saturn opposition Uranus, indicative of prolonged atmospheric conditions with the potential to affect an entire region for months on end because of the slow movement of these planets. That aspect also signified cold temperatures, sudden and extreme atmospheric disturbances, and more intense downfall. Saturn also opposed Pluto (windy storm). The triple combination—activated by the transiting Moon in Leo semisquare ingress and lunar phase Uranus conjunct Pluto (extreme weather), sesquisquare lunar phase and ingress Saturn, square the lunar phase Ascendant, and opposition the lunar phase Aquarius planets (the lunar phase occurred about five hours before the snow began, when the transiting Moon was square the lunar phase Ascendant)—resulted in an unexpected storm that crippled Chicago, the major transportation hub of the Midwest.

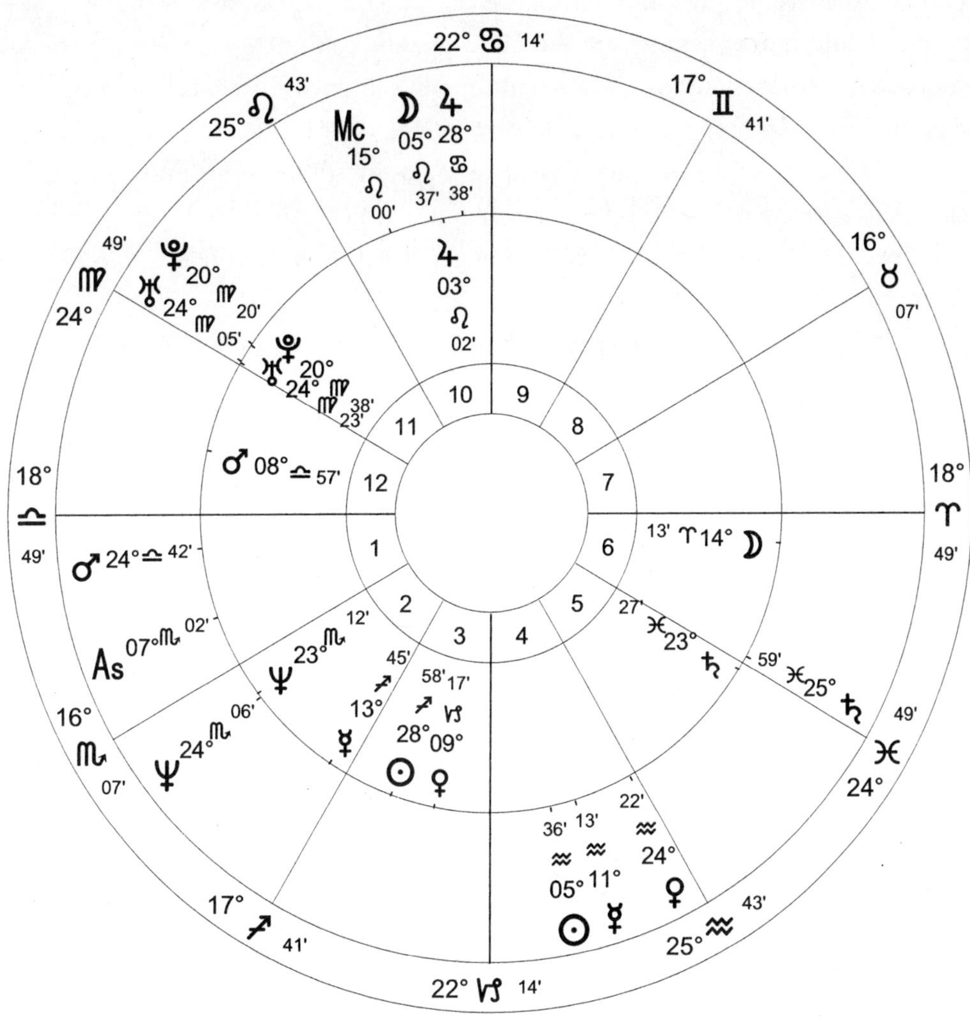

Snow 4

INNER WHEEL=*Winter Ingress / Natal Chart / December 21, 1966, Wed / 1:29 am CST +6:00*
Chicago, IL / 41°N51' 087°W39' / Geocentric / Tropical / Placidus / Mean Node

OUTER WHEEL=*Lunar Phase / Natal Chart / January 26, 1967, Thu / 0:41 am CST +6:00*
Chicago, IL / 41°N51' 087°W39' / Geocentric / Tropical / Placidus / Mean Node

Additional factors present in the ingress chart were Uranus sextile Neptune (unusual weather), and Jupiter semisquare the Pluto-Uranus conjunction to amplify the high winds and stormy conditions. Mars semisquare Neptune indicated the moist air that met the Saturn-Uranus cold front. However, it was the previous lunar phase (January 18, 1967) that showed the strong flow of warm, moist, southerly air into the area. That chart had a Mercury-Jupiter opposition spanning the Midheaven/IC. On January 24, 1967, two days before the snow began falling, Chicago was an unseasonable 65 F with thunderstorms.

The full fury of the storm was apparent by mid-morning on January 26, 1967, when the transiting Moon opposed lunar phase Mercury, and by 6:00 pm, Chicagoans had no doubt they were in the midst of a major storm when the transiting Moon opposed the lunar phase IC. It wasn't until the next morning, when the transiting Moon was separating from an opposition to the lunar phase Venus-Neptune square, that snow stopped falling.

The ingress chart alone was not enough to indicate this major snowstorm in Chicago, despite the trines, sextiles, and inconjuncts from the outer planets to the meridian. That would happen with the lunar phase of January 26, 1967 (outer wheel of Snow 4).

On its own, the lunar phase chart warned of a major snowfall with Venus in icy Aquarius in a nearly exact square to Neptune in Scorpio. That square, an indicator of heavy precipitation, was amplified by Venus parallel the Ascendant and Neptune parallel the IC. The lunar eclipse of October 29, 1966, at 5 Taurus was conjunct the Descendant, and the November 12, 1966, solar eclipse at 20 Scorpio was square the IC.

Mercury, a major player in the storm, was also in Aquarius. It was conjunct the IC, warning of a potential storm emergency due to gusty winds, as did its parallel to the Sun and sesquisquare to Uranus. The Sun-Uranus-Pluto sesquisquare indicated a cold front, overcast skies, wind gusts, and snow, with the Uranus-Pluto conjunction activating the meridian through a sextile/trine; it amplified the cold temperatures already promised by Aquarius.

Aspects to the ingress chart that compounded the potential of the lunar phase chart were lunar phase Venus sesquisquare ingress Mars and semisquare ingress Venus, lunar phase Jupiter sesquisquare ingress Mercury, and the lunar phase Sun opposition ingress Jupiter.

The outer planetary aspects of the ingress chart were still in force, being kicked off by transiting Mars at 24 Libra—square Jupiter, inconjunct Saturn, and semisextile Uranus—and the aspects made by the faster-moving Mercury and Venus in Aquarius. Mercury's main function was to trigger the outer planetary trio with its exact semisquare to Saturn, while Venus augmented the downfall with its square to Neptune.

But it was the transiting Moon in Leo square the lunar phase Ascendant and aspecting the Saturn-Uranus-Pluto (semisquare and sesquisquare) in both charts that set things in motion. The lunar phase also occurred in the fourth and tenth houses of the ingress chart, further bringing Chicago into focus as the recipient of a major winter storm. The Moon continued to travel throughout the storm, opposing lunar phase Mercury, the lunar phase IC, and, finally, lunar phase Venus, when it became void-of-course and the snow ended.

A close examination of the charts involved in this storm easily explains why such storms rarely occur. Necessary factors are major outer planetary configurations combined with transiting planets in the exact positions necessary to trigger them from many perspectives. Every planet, nearly all in hard aspect, were involved in and necessary to the formation of a storm of this magnitude.

Other aspects to the angles that made Chicago the storm's target were lunar phase Jupiter contraparallel the ingress IC, ingress and lunar phase Neptune parallel the lunar phase IC, ingress Pluto contraparallel the ingress IC, and lunar phase Venus parallel the lunar phase Ascendant. Most important, lunar phase Saturn was sesquisquare the lunar phase Ascendant, a point that also was semisquare Uranus and Pluto. Ingress Mercury was parallel the ingress IC, and ingress Saturn, Pluto, Uranus, and Neptune aspected the same angle.

ICE STORM, JANUARY 26–28, 2009

Louisville, Kentucky, was at the center of an ice storm that left 700,000 residents without power and another 200,000 without water as a result of felled trees and downed power lines. Power was not completely restored until ten days after the storm. Precipitation began as freezing drizzle and rain and then changed to sleet and snow in some of the affected areas. Ice more than an inch thick was reported in some locations.

The December 21, 2008, winter ingress chart (inner wheel of Snow 5) had prime indicators for heavy precipitation, storms, and freezing conditions: Neptune semisquare/sesquisquare the meridian, Uranus square the horizon, and Mars in a T-square with Saturn and Uranus.

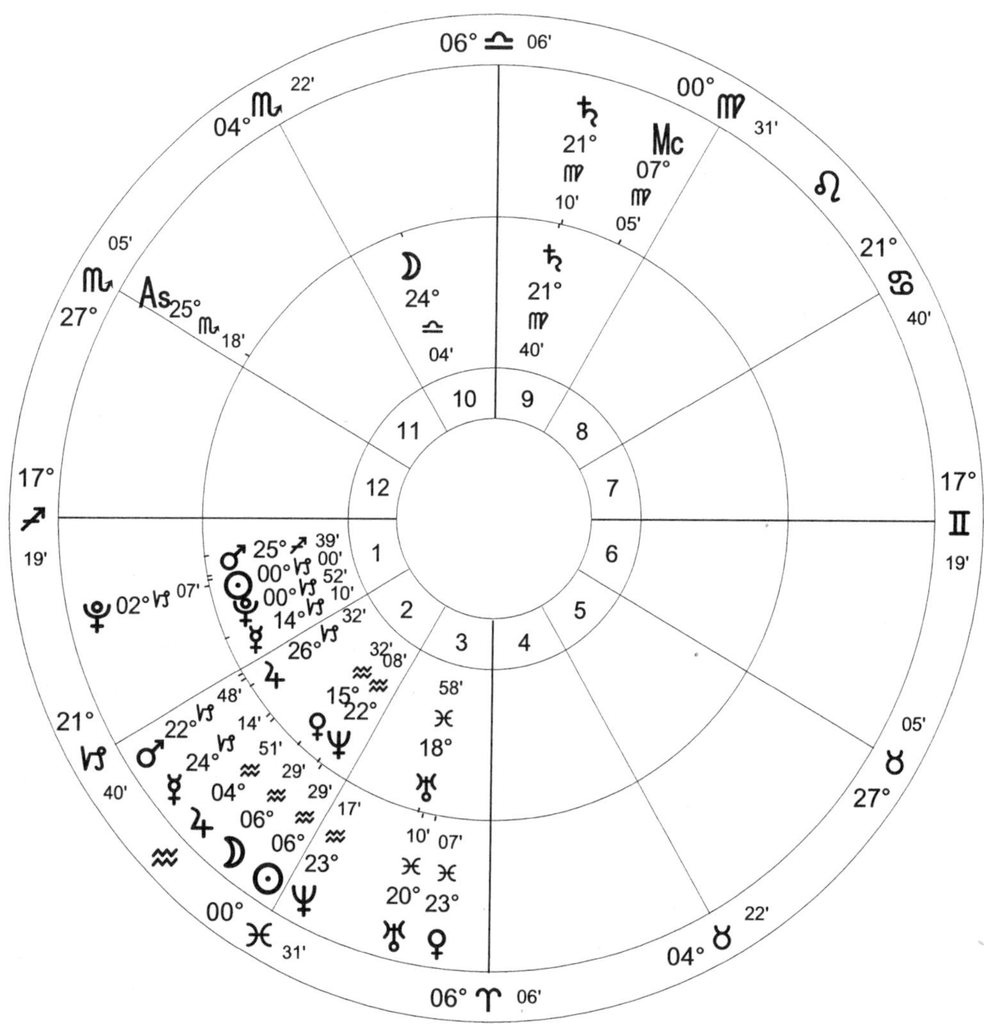

Snow 5

INNER WHEEL=*Winter Ingress / Natal Chart / December 21, 2008, Sun / 7:03:46 am EST +5:00*
Louisville, KY / 38°N15'15" 085°W45'34" / Geocentric / Tropical / Placidus / Mean Node

OUTER WHEEL=*Lunar Phase / Natal Chart / January 26, 2009, Mon / 2:55:17 am EST +5:00*
Louisville, KY / 38°N15'15" 085°W45'34" / Geocentric / Tropical / Placidus / Mean Node

In the January 26, 2009, lunar phase chart (same date as when the storm began—outer wheel of Snow 5), retrograde Mercury was conjunct Mars (an aspect of wind and cold that would become exact during the course of the storm) and semisquare/sesquisquare the meridian. Lunar phase Venus was conjunct Uranus (cold, drizzle), with both planets trine/sextile the horizon and opposition Saturn (heavy precipitation). As in the ingress chart, Neptune was active, this time in a square to the horizon.

Lunar phase Pluto (wind, cold) was square the ingress meridian and semisquare Venus (intense precipitation and wind), and lunar phase Venus was square ingress Mars (storm, precipitation), another aspect that became exact during the course of the storm. The storm persisted as the transiting Moon traveled through Aquarius, aspecting every significant planet in the ingress and lunar phase charts; three to four inches of snow accumulated on January 28 after the Moon entered Pisces.

A big-picture view of these charts shows a strong contrast of warmer and colder air. Venus, Mars, and Neptune are warm, while Uranus, Saturn, and retrograde Mercury are cold. Notice how the cold and warm planets aspected each other, so it was as though the atmosphere couldn't decide whether there should be snow, sleet, or freezing rain!

Freezing Rain and Sleet, February 14, 1990

An excellent example of the transition from rain, to sleet, to freezing rain, to snow is the weather event of February 14, 1990. On that date, Chicago (latitude 41N52) received nine inches of snow, while southern Illinois (38N23) was deluged with rain. In between, Champaign, Illinois (40N07), experienced one of the worst ice (and sleet) storms in its history. Freezing rain caused an accumulation of an inch and a half of ice on every surface, including power lines, and most citizens were without electricity for a week. Damage reached $9 million.

Here again there is a significant mixed indication of cold and warm influences, so moisture could take the form of freezing rain, sleet, and snow.

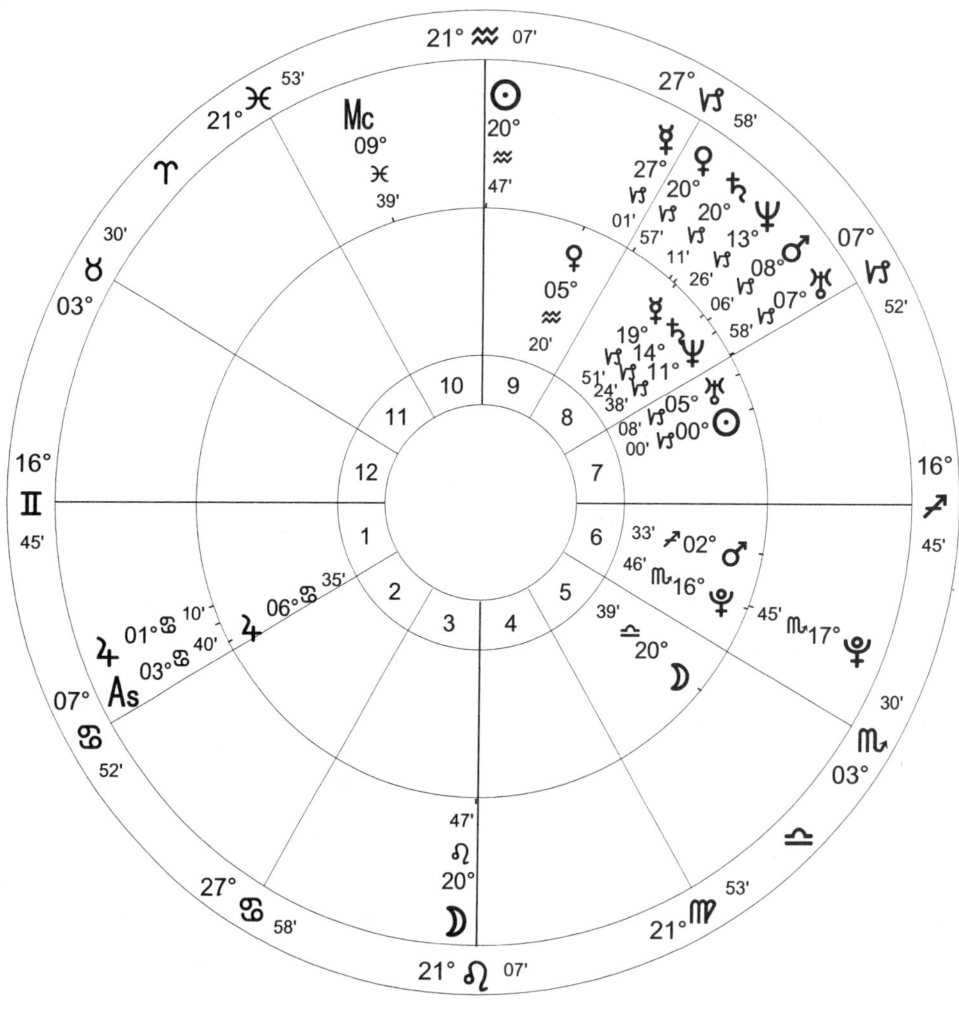

Snow 6

Inner Wheel=*Winter Ingress / Natal Chart / December 21, 1989, Thu / 3:23 pm CST +6:00*
Chicago, IL / 41°N51' 087°W39' / Geocentric / Tropical / Placidus / Mean Node

Outer Wheel=*Lunar Phase / Natal Chart / February 9, 1990, Fri / 1:17 pm CST +6:00*
Chicago, IL / 41°N51' 087°W39' / Geocentric / Tropical / Placidus / Mean Node

The planetary aspects, with the exception of those to the angles, are identical for all three locations. Both the ingress and February 9, 1990, lunar phase charts (Snow 6) show the conflicting message of warm and cold aspects. At the ingress, Sun-Mars, Sun-Pluto, Mercury-Jupiter, and several Venus aspects indicated warmer weather, while the classic cold weather planets, Saturn and Uranus, were strongly aspected. This mixture was evident at the lunar phase as well. There, Sun-Venus, Sun-Mars, and Venus-Pluto aspects indicated warmth, while aspects to Saturn and Uranus promised cold temperatures.

Heavy moisture in combination with falling temperatures was the expectation in the ingress chart, with Mercury parallel Uranus, Jupiter contraparallel and opposition Uranus, and Saturn conjunct and parallel Neptune. Uranus was semisquare Pluto, suggestive of sleet and a storm emergency centering around power and communication lines.

The lunar phase chart had the key element of stationary direct Venus conjunct Saturn, an indicator of prolonged, heavy precipitation that can take the form of rain, sleet, or snow. Mars conjunct and parallel Uranus reinforced the potential for sleet and freezing rain at the appropriate latitude. Other factors in the lunar phase chart were Saturn parallel and sextile Pluto, Jupiter contraparallel Uranus and sesquisquare Pluto, and Mars contraparallel Jupiter. Mercury parallel Saturn and Neptune ensured leaden, overcast skies and precipitation.

There were significant connections between the lunar phase and ingress charts, many of which repeated and reinforced aspects in the lunar phase chart, intensifying the precipitation. Ingress Saturn and lunar phase Neptune formed a conjunction, as did ingress Mercury and lunar phase Saturn, adding fuel to the parallels in the lunar phase chart. In addition to conjoining lunar phase Uranus, lunar phase Mars made the same aspect to ingress Neptune. Air flow temperatures varied as ingress Mercury made a conjunction to the lunar phase Venus-Saturn conjunction. Additional warmth was provided by the ingress Sun–lunar phase Jupiter opposition and lunar phase Sun–ingress Jupiter sesquisquare.

On February 14, the storm was triggered by the transiting Moon in Libra square the lunar phase Venus-Saturn conjunction and ingress Mercury. Precipitation (sleet, rain, snow) arrived with the transiting Sun at 25 Aquarius semisquare ingress Saturn and Neptune. Sleet and freezing rain were indicated by two "double" February 14 transits to the ingress chart—transiting Mercury at 4 Aquarius parallel ingress Mars and transiting

Mars at 11 Capricorn parallel ingress Mercury—and transiting Mars parallel ingress and lunar phase Uranus. Transiting planetary aspects that day were Sun semisquare Uranus (sleet or snow, cold and damp), Mars exact contraparallel Jupiter (storm), Mars exact parallel Uranus (sleet), and Jupiter exact parallel Uranus (intense cold front and high precipitation).

There were numerous lunar phase planetary aspects to the ingress chart that brought this storm to Chicago, Champaign, and southern Illinois. In all locations, lunar phase Pluto was square the ingress meridian, lunar phase Mars and Uranus were sesquisquare/semisquare the ingress meridian, lunar phase Mercury was sesquisquare/semisquare the ingress horizon, and the lunar phase Sun was conjunct the ingress Midheaven. The predominant transits to all three ingress charts were transiting Venus and Saturn at 21 Capricorn inconjunct the IC, bringing the key ingress seventh house Venus-Saturn conjunction (as well as ingress Mercury) to the meridian. As in other major storms, the ingress Sun contacted an angle, in this case a conjunction to the lunar phase Descendant.

There is one dominant factor in the lunar phase charts that explains why the storm brought snow to Chicago (Ascendant 3 Cancer 42), rain to southern Illinois (Ascendant 0 Cancer 32), and sleet and freezing rain to Champaign (Ascendant 1 Cancer 52). Jupiter, a classic indicator of warmer weather, was conjunct the Ascendant and affected the temperature of the three locations. In the Chicago chart, it was behind the Ascendant, a placement that has a diminished effect on the point of observation. In Champaign, it was on the Ascendant, acting as the transition point between colder (Chicago) and warmer (southern Illinois) temperatures. In southern Illinois, it was in front of the Ascendant, bringing warmer temperatures to that location.

CHAPTER 10

Thunderstorms, Hail, and Tornadoes

One of nature's most spectacular displays of energy is the thunderstorm. These storms differ from other storms in that they are convective, an atmospheric phenomenon that occurs when warm air rises, cooling on ascent, and then descends, all within the thunderstorm. As this cycle is repeated, the thunderstorm increases in size. In order to form, a thunderstorm needs unstable air: warm air near the earth and cold air in the higher atmosphere.

A thunderstorm begins to form when warm, moist air rises from the earth, creating an updraft. As the air cools and condenses, a cloud forms; when the water droplets reach the same temperature as their higher-altitude environment, they begin falling, creating a downdraft. The faster the air rises within the storm structure, the more intense the thunderstorm. Updrafts continue to feed the storm until they are eventually overcome by the cooler downdrafts, and the storm dissipates.

Unlike a small-scale thunderstorm (about fifteen miles in diameter with a life span of about thirty minutes), a super cell thunderstorm, which often produces hail and tornadoes, can last for hours and travel more than 300 miles during its life. Squall lines, such as the derecho, are also longer-life thunderstorms that can travel great distances.

Thunderstorms are common in spring and summer, but occur less frequently in autumn and winter. The strongest storms are generally in the spring months, when there is the greatest contrast of warm surface air and cold upper air, which results in the necessary atmospheric instability. The requirement for warm surface air also explains the timing of most thunderstorms during the summer months. During the day, the heat builds, reaching a temperature of greatest contrast with the upper atmosphere in the late afternoon and early evening.

Squalls, such as the derecho, are a line of longer-life thunderstorms, sometimes more than 100 miles in length, that can travel great distances. They often form ahead of an advancing cold front, creating strong winds and sometimes hail and tornadoes.

A thunderstorm needs the same planetary moisture components as a snowstorm (Venus, Neptune, or planets in water signs), and an infusion of cold air from Saturn or Uranus and warm air from Mars or Jupiter. The wind characteristic of thunderstorms, especially super cells and squalls, comes from Mercury, Mars, and Pluto. A likely combination of cold and warm air in a thunderstorm is, for example, Mars (heat) in hard or neutral aspect with Uranus (cold). The presence of Saturn or Neptune represents the drop in barometric pressure that accompanies a thunderstorm, and Mercury-Uranus and Mars-Uranus indicate hail.

THUNDERSTORM, JUNE 17, 2013

This thunderstorm is typical of those that occur throughout the spring and summer across North America, generally popping up in the late afternoon or evening. There were no tornadoes reported with this storm, although some areas saw hail and wind gusts as high as 65 mph. Rainfall amounts ranged from half an inch to about two inches, depending on the location. Lubbock was at the center of the thunderstorm band that extended northwest to southeast through that area of Texas.

The March 20, 2013, spring ingress and the June 16, 2013, lunar phase charts (inner and middle wheels of Thunderstorms 1) are a good example of how an ingress chart with two severe thunderstorm/tornado signatures (Mercury conjunction Neptune and Mars conjunction Uranus) can indicate a lesser-strength thunderstorm without tornadoes. Despite strong connections between the ingress and lunar phase charts (lunar phase horizon square ingress meridan), there were no planetary aspects or aspects to the horizon and meridian to trigger the signatures during this lunar phase.

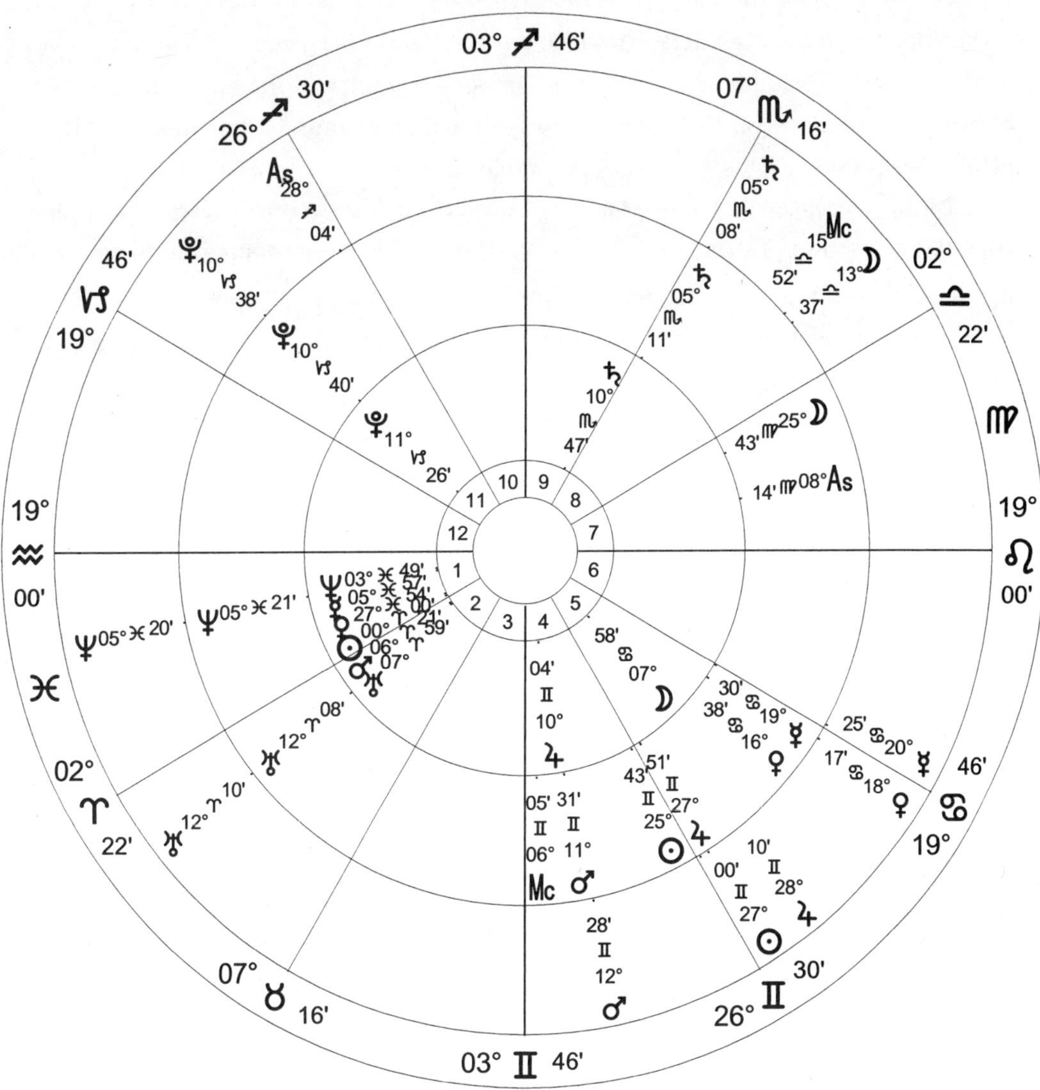

Thunderstorms 1

Inner Wheel=*Spring Ingress / Natal Chart / March 20, 2013, Wed / 6:01:55 am CDT +5:00*
Lubbock, TX / 33°N34'40" 101°W51'17" / Geocentric / Tropical / Placidus / Mean Node

Middle Wheel=*Lunar Phase / Natal Chart / June 16, 2013, Sun / 12:23:41 pm CDT +5:00*
Lubbock, TX / 33°N34'40" 101°W51'17" / Geocentric / Tropical / Placidus / Mean Node

Outer Wheel=*Thunderstorms / Natal Chart / June 17, 2013, Mon / 9:00 pm CDT +5:00*
Lubbock, TX / 33°N34'40" 101°W51'17" / Geocentric / Tropical / Placidus / Mean Node

Nevertheless, the Mercury-Neptune-Descendant conjunction in water signs was a strong player in this thunderstorm from a precipitation perspective. Lunar phase Saturn (low pressure, cloudy skies) was trine the conjunction, and the lunar phase Venus-Mercury conjunction (precipitation) was sesquisquare the same conjunction. All of these planets were in water signs. Hail was indicated by lunar phase Mars sextile the lunar phase Mars-Uranus conjunction, with Mars separating from a conjunction to ingress Jupiter (thunderstorm) and in a wide conjunction with the lunar phase Midheaven. Ingress Saturn was activated through a sesquisquare from the lunar phase Sun-Jupiter conjunction.

The transiting planetary positions (outer wheel of Thunderstorms 1) were nearly the same as the lunar phase because this thunderstorm occurred the next day. But in that time, Venus had advanced to conjunct lunar phase Mercury, doubling the precipitation indication, and Mercury had moved into an exact sesquisquare with lunar phase Neptune. All of these aspects were related to rain, but they weren't strong enough (a sesquisquare rather than a square, opposition, or conjunction to Mercury-Neptune) to indicate a tornado-producing super cell. Had lunar phase and transiting Mars been in early Gemini and square the Mercury-Neptune conjunction, Lubbock would have seen a tornado. That happened in early June when Mars was at 3 Gemini.

DERECHO, MAY 8, 2009

A derecho is a fast-moving, straight-line band of thunderstorms with high winds that can travel hundreds of miles; tornadoes often occur with this kind of squall. They are characterized by a bow echo radar image (the image looks like the fully extended string of a bow).

The derecho of May 8, 2009, which had its genesis in northeastern Colorado/northwestern Kansas, traveled to western West Virginia before dissipating more than 1,000 miles later. Hundreds of trees toppled, buildings had significant damage, power outages were extensive, flash flooding occurred in several areas, and among the many tornadoes were six that were rated EF-2 and one rated EF-3. From genesis to dissipation, the storm lived for more than twenty-five hours, an extremely long time frame for a derecho, which prompted some meteorologists to call it a Super Derecho. This storm continues to be studied by meteorologists as they strive to understand its development and intensity; astrometeorology can provide at least some of the answers.

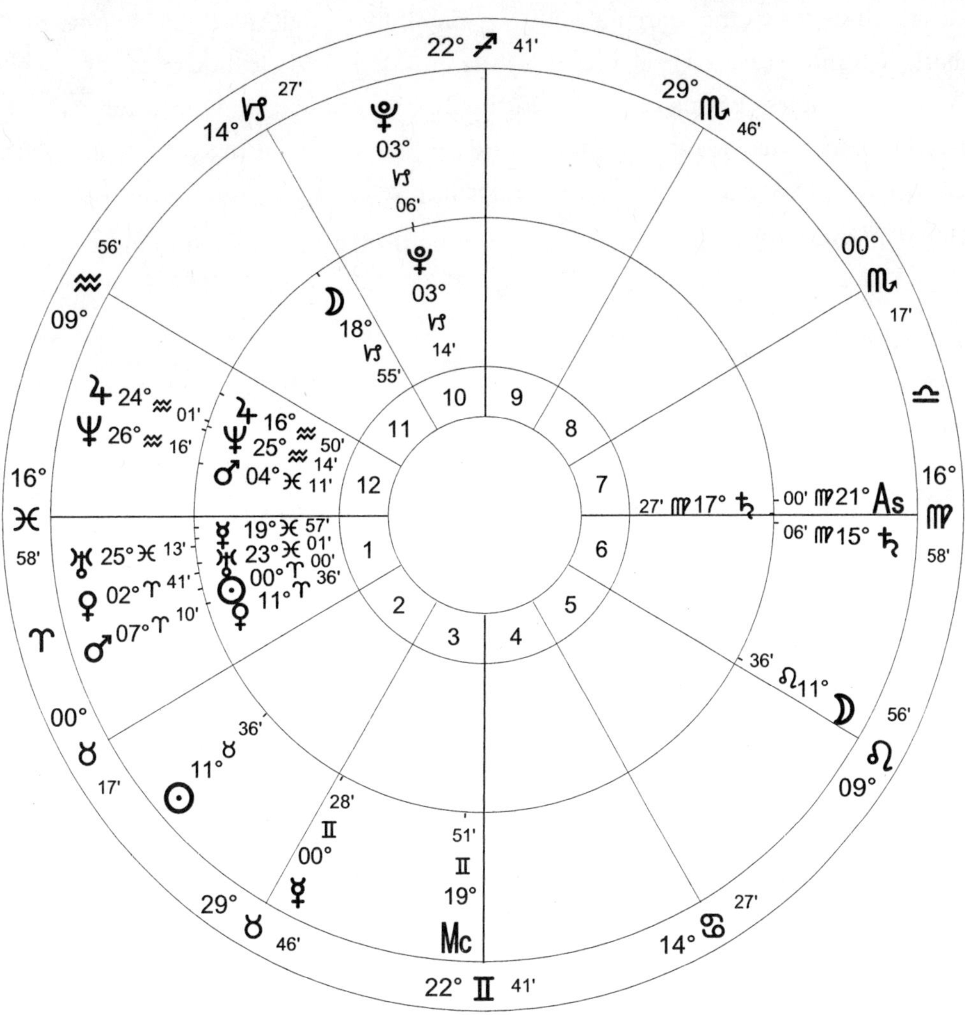

Thunderstorms 2
INNER WHEEL=*Spring Ingress / Natal Chart / March 20, 2009, Fri / 6:43:38 am CDT +5:00*
38°N00' 092°W00' / Geocentric / Tropical / Placidus / Mean Node

OUTER WHEEL=*Lunar Phase / Natal Chart / May 1, 2009, Fri / 3:44:13 pm CDT +5:00*
38°N00' 092°W00' / Geocentric / Tropical / Placidus / Mean Node

The chart shows the March 20, 2009, spring ingress and May 1, 2009, lunar phase charts (Thunderstorms 2) calculated for the longitude and latitude of the area where the bow was largest, having reached maturity. To better understand the dynamics of this derecho, here is a table of the Ascendants and Midheavens of the ingress and lunar phase charts across the area through which the storm traveled (101W is Colorado/Kansas and 81W is West Virginia).

Longitude	40N	39N	38N	38N	38N	38N	38N
Latitude	101W	101W	98W	95W	90W	85W	81W
Ingress ASC	1 PI	0 PI	5 PI	13 PI	19 PI	29 PI	23 PI
Ingress MC	25 SG	13 SG	16 SG	20 SG	24 SG	29SG	2 CP
Lunar Phase ASC	14 VI	14 VI	16 VI	18 VI	22 VI	26 VI	29 VI
Lunar Phase MC	11 GE	11 GE	14 GE	17 GE	22 GE	26 GE	29 GE

One of the most striking components of these charts is ingress Saturn at 17 Virgo and lunar phase Saturn at 15 Virgo. Compare these with the listed Ascendants and Midheavens and consider that almost immediately after genesis, the storm began to take a southeasterly direction. Once the storm had reached a longitude and latitude where lunar phase Saturn was in the twelfth house instead of the first, the storm switched to an easterly direction, which it maintained until dissipation. So the lower pressure represented by Saturn guided the storm's direction.

The longitude and latitude where Saturn was conjunct the lunar phase Ascendant and square the lunar phase Midheaven was where the storm reached maximum intensity with a mature bow. At the same location, Saturn was square the ingress Midheaven. Using longitude and latitude and one key planet, you can thus identify the strongest point of a large storm.

Now rewind to the point of genesis, which was rather unremarkable given what evolved during the twenty-five hours that followed. The key planet here, and one that would alert an astrometeorologist that there was far more to come, was lunar phase Mercury—and, by the date of the event, transiting stationary retrograde Mercury—square the ingress Ascendant at 39–40N latitude, 101W longitude. This planet of wind is at its worst when stationing, and secondarily during its retrograde period. So when it forms a hard aspect to a lunar phase or ingress horizon or meridian, a weather event is more than likely, as is one that involves high wind velocity.

The possibility for strong thunderstorms and tornadoes is apparent in the ingress chart, and at the location of the derecho's greatest intensity. There, the ingress Mercury-Uranus conjunction opposition Saturn (thunderstorms, tornadoes, low pressure, high pressure, hail, wind) was conjunct the ingress Ascendant and square the ingress Midheaven. That configuration was sextile/trine the ingress Moon, linking all four planets. This is an important factor because Mercury turned retrograde sesquisquare the ingress Moon, providing the necessary link between the lunar phase and ingress charts at the longitude and latitude of the storm's genesis.

Other active aspects were transiting Mars conjunction ingress Venus (precipitation, storms, warm and moist air), transiting Mars semisquare ingress and lunar Neptune (tornadoes, storms, precipitation, warm and moist air), and transiting Sun sesquisquare ingress and lunar phase Pluto (wind, storms). The declinations were Venus parallel Mars (precipitation, storms) and Venus and Mars contraparallel Uranus (cold air, warm air, storms, precipitation, hail).

A close look at the transiting Moon reveals why the storm began and ended when it did. At the onset, the transiting Moon was at 5 Scorpio, sesquisquare ingress Mercury opposition Saturn conjunct the horizon. This was the triggering aspect. The storm had mostly dissipated in West Virginia when the Moon was at 17 Scorpio sextile ingress Saturn and sesquisquare 1 Gemini, where Mercury had stationed retrograde the day before. Just before midnight, the time of the next lunar phase, high winds damaged several buildings near the West Virginia/North Carolina border.

HAIL, SEPTEMBER 5, 1898

Some thunderstorms are accompanied by hail, a condition that results when ice crystals in the upper atmosphere have no opportunity to melt. The ice crystals continue to grow as the updrafts in the thunderstorm keep them aloft and very cool water continues to freeze on them, adding to their size. The faster the updraft, the bigger the hailstones. Generally, hail forms in only one part of the storm, which is why only selected smaller areas on Earth are bombarded with the stones.

Hailstones usually are no more than about one-quarter inch in diameter, although some as large or larger than baseballs have been recorded; the largest ever recorded was eight inches in diameter.

Hail generally falls for only a few minutes, during which time it can cause extensive damage to crops, vehicles, and buildings. In rare cases, hail has fallen for an hour or more, leaving inches of the icy stones on the ground for extensive periods of time. The largest hailstones cause much destruction, falling at as much as 100 mph and weighing more than a pound.

The most prevalent aspects during a hailstorm are Mercury in hard or neutral aspect to Mars or Pluto, Mercury in any aspect to Uranus, and Mars in hard or neutral aspect to Uranus. These must, of course, be accompanied by traditional thunderstorm aspects.

One of the worst hailstorms on record occurred in Nodaway County in northwest Missouri on September 5, 1898. Hail remained on the ground in some places for fifty-two days, and ice-clogged farm fields were unworkable for two weeks.

The June 21, 1898, summer ingress chart (inner wheel of Thunderstorms 3) had the necessary thunderstorm-breeding aspects: Sun square Jupiter (warmer and thunderstorms), Mercury conjunct and parallel Neptune (thunderstorms, tornadoes), Venus trine Uranus (warm and cold air), Mars semisquare Neptune (rising, warm, moist air), and Saturn parallel Uranus (atmospheric disturbance). Mercury and Neptune, strongly placed, were conjunct the Ascendant, waiting to be activated by future transits and a lunar phase chart. The Moon, semisextile the IC, was parallel Venus, contraparallel and sesquisquare Saturn, contraparallel Uranus, and semisextile Neptune.

The focus of the August 31, 1898, lunar phase chart (middle wheel of Thunderstorms 3) was thunderstorms, hail, and wind. Planetary configurations for precipitation were the Sun semisquare Venus, Venus semisquare Saturn, Venus trine Neptune, and Mars conjunct Neptune. The Sun square Saturn and retrograde Mercury square Pluto signaled thunderstorms and hail. Windy conditions were indicated by retrograde Mercury square Pluto (also hail), Mars inconjunct Uranus, and Jupiter semisquare Uranus. Venus represented warm, moist air, and Jupiter semisquare Uranus, Mars inconjunct Uranus, Sun square Saturn, and Saturn parallel Uranus indicated cold air.

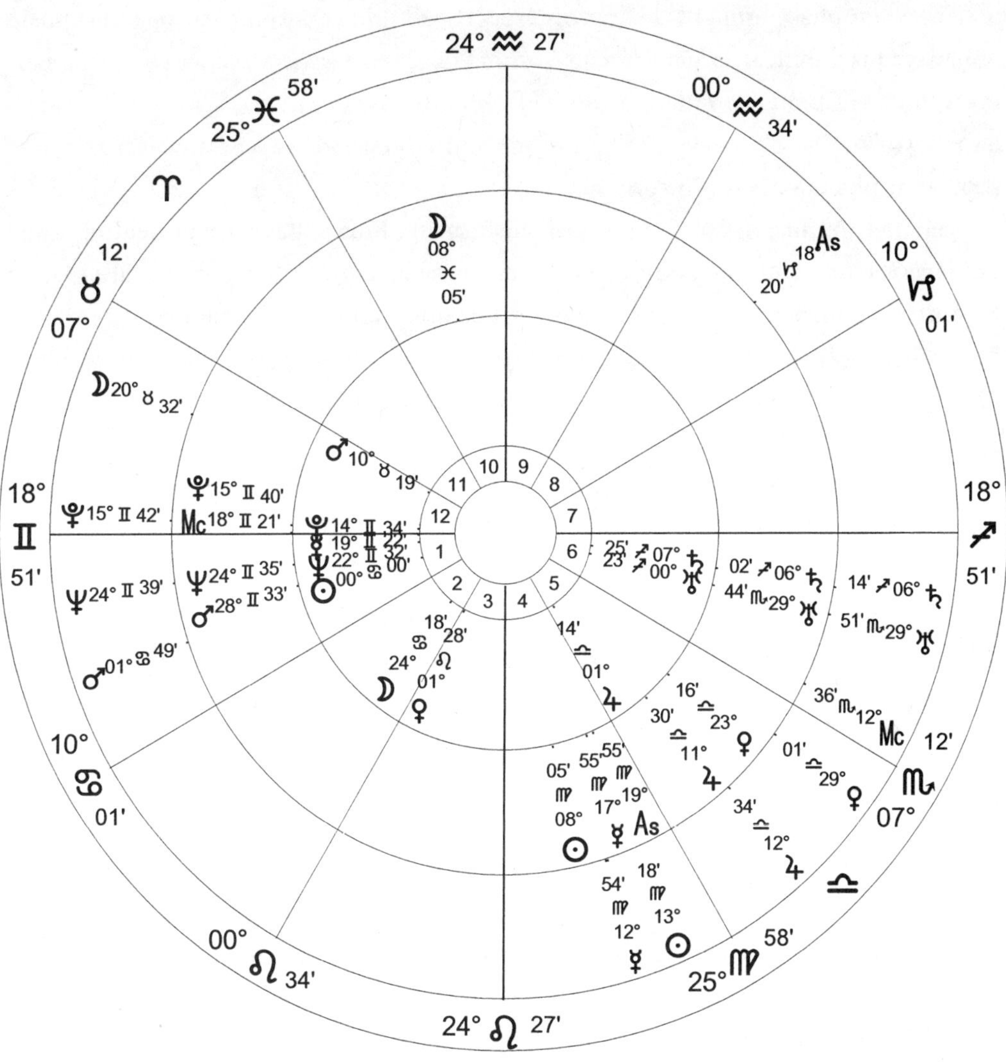

Thunderstorms 3

Inner Wheel=*Spring Ingress / Natal Chart / June 21, 1898 NS, Tue / 4:08 am CST +6:00*
Nodaway County, MO / 40°N21' 094°W52' / Geocentric / Tropical / Placidus / Mean Node

Middle Wheel=*Lunar Phase / Natal Chart / August 31, 1898 NS, Wed / 6:50 am CST +6:00*
Nodaway County, MO / 40°N21' 094°W52' / Geocentric / Tropical / Placidus / Mean Node

Outer Wheel=*Hail / Natal Chart / September 5, 1898 NS, Mon / 4:00 pm CST +6:00*
Nodaway County, MO / 40°N21' 094°W52' / Geocentric / Tropical / Placidus / Mean Node

The lunar phase brought the ingress aspects and others into force. The lunar phase Sun was square ingress Saturn (storms), lunar phase retrograde Mercury was square ingress Pluto (heat, high winds, storms, hail), lunar phase Venus was semisquare ingress Saturn (low pressure, heavy precipitation) and trine ingress Neptune (warm, moist air), lunar phase Mars was inconjunct ingress Uranus (warm and cold air), and Saturn was parallel Uranus. Aspects in both the ingress and lunar phase charts were repeated between the two with the addition of lunar phase retrograde Mercury square ingress Mercury conjunct the Ascendant. The ingress Sun, Mercury, and Neptune were contraparallel the lunar phase IC, lunar phase Jupiter was semisquare the ingress IC, lunar phase Mars and Neptune were parallel the ingress Ascendant, and lunar phase Pluto was parallel the ingress IC.

On the date of the storm, September 5, 1898 (outer wheel of Thunderstorms 3), the ingress chart was activated by nearly every transiting planet. Thunderstorms were promised with the transiting Sun and Mercury at 13 Virgo square ingress Pluto, transiting Venus at 29 Libra sesquisquare ingress Pluto, transiting Neptune at 24 Gemini semisquare ingress Mars, and transiting Mars at 1 Cancer square ingress Jupiter and parallel the ingress Sun. One of the characteristic aspects of hail, Mercury square Pluto, was present, and the transiting Moon in Taurus served as a trigger with its parallel to the ingress Sun, Mercury, Venus, and Neptune.

Thunderstorms were reflected in the aspects to the lunar phase chart as well: transiting Venus contraparallel lunar phase Pluto and transiting Mercury contraparallel lunar phase Jupiter. Mercury, already strong from its placement in the ingress chart, was at 0 degrees north declination in the lunar phase chart and retrograde. Transiting Mars and Neptune were parallel the lunar phase Ascendant, which was contraparallel transiting Jupiter, and transiting Venus was contraparallel the lunar phase IC.

Transiting planetary aspects reflected thunderstorm activity with hail. Transiting retrograde Mercury at 13 Virgo was square transiting Pluto at 15 Gemini and semisextile transiting Jupiter at 12 Libra. The transiting Sun at 13 Virgo was conjunct transiting retrograde Mercury, semisextile transiting Jupiter, semisquare transiting Venus, and square transiting Pluto. Transiting Venus was approaching a trine to transiting Mars and semisextile transiting Uranus. Parallels and contraparallels were numerous: Moon-Neptune, Moon-Mars, Mercury-Jupiter, Venus-Pluto, and Saturn-Uranus.

Note again that Mercury, Mars, and Pluto, planets associated with hail, all aspected the lunar phase meridian. This, in effect, brought these planets together to intensify the hailstorm at that location. Even more telling were lunar phase Mars and Neptune parallel the ingress Ascendant. Nodaway County was also highlighted because the lunar phase IC was opposition the ingress Ascendant and the lunar phase Ascendant was square the ingress Ascendant.

The long-term influence of Saturn aspecting Uranus and Mercury conjunction the Ascendant were the overriding factors that kept temperatures cool enough to hamper quick melting of the hailstones. Later in the month, Mercury stationed direct in a square to Saturn. As this aspect separated and Mercury was no longer retrograde, temperatures began to warm. However, it wasn't until the autumn ingress of September 22, 1898, that the weather pattern changed. At that time, Mars was approaching a square to Jupiter, which would have accelerated the melting of the hailstones during succeeding weeks. Even so, there was enough hail left a month later for the residents to use it to make ice cream.

Tornadoes

A tornado is one of nature's most frightening and destructive forces. It is defined by one distinguishing feature: a whirling, tapering column of high-velocity wind that extends down from a cumulonimbus (thunderstorm) cloud. A funnel cloud, which becomes a tornado when it touches ground, can range in size and color from one that is thin, rope-like, and grayish white to a wide, thick, black, whirling mass of dirt and debris. Tornadoes rotate in a counterclockwise direction.

The destructive power of a tornado comes from its high winds, which damage objects in its path in several ways: a blast effect from wind speed alone; the explosion of buildings and other objects because of reduced external atmospheric pressure; the lifting of objects, including roofs, because of a difference in air pressure; and flying debris. Some tornadoes produce multiple funnels.

A tornado is initially visible because of the condensation of water vapor around the whirling cloud. Later, as it gathers dust and debris, the central funnel often is hidden inside a dark column.

The exact cause of tornadoes is not completely understood, although extensive research is ongoing and much has been learned by tornado chasers. However, there are two known conditions necessary for a tornado to form: instability and wind shear. Instability results when there is warm, moist air in the lower atmosphere/surface and much colder air in the upper atmosphere; these are the same conditions necessary for a thunderstorm. Wind shear is defined as wind direction changing and increasing with height; for example, a 10 mph southerly wind at the surface could become a westerly or southwesterly wind of 60 mph at 5,000 feet. These conditions generally exist only ahead of a cold front and low-pressure system.

Certain areas of the United States are prone to tornadoes at various times of the year, primarily because of temperature contrast. The Gulf states can experience tornadoes in the winter months. As temperatures warm in March and April, tornadoes form farther northward across the Plains states, eventually moving into the Upper Midwest and Great Lakes states in May and June and remaining active through September. In November and December when the Plains states experience weather too cold for thunderstorms, tornadoes again move south.

Oklahoma sees the most tornadoes per square mile, and more than half the tornadoes each year occur in the central and southern Plains states. An area often referred to as Tornado Alley runs from Texas across Oklahoma, Kansas, and Nebraska to the Dakotas. Greatest activity occurs along the 97W and 98W meridians. Increased tornado activity also occurs in a curving line from west Texas, to Missouri, to Iowa, Illinois, Indiana, and Ohio. Dixie Alley extends from southeast Texas across the southeastern United States. Tornadoes also are occasionally seen in central New England.

Because tornadoes most often occur in the warmer part of the day when conditions are favorable for thunderstorm development, tornado activity is more prevalent in the afternoon and early evening, especially in the Midwest and Plains states. But they can occur at any time of day.

Most tornadoes move in a southwest to northeast direction at varying speeds, with some following a more west to east path. Tornado strength is measured according to the Enhanced Fujita scale (EF scale):

Scale	Wind Speed
EF0	65–85 mph
EF1	86–110 mph
EF2	111–135 mph
EF3	136–165 mph
EF4	166–200 mph
EF5	greater than 200 mph

For a tornado to form, classic thunderstorm aspects are a must, including, for example, Mars opposition Saturn (cold and warm fronts colliding) and a strong presence of Venus and/or Neptune (moisture). With the chief component of tornadoes being wind, Mercury, Mars, or Pluto must be prominent in the chart, along with at least one other planet to amplify it. A tornado rarely, if ever, results without Mercury or Mars in hard aspect to Uranus or Neptune, a necessity for the characteristic counterclockwise winds. It also is rare for a tornado of any size, especially an EF 3, 4, or 5, to form without a hard aspect to Pluto, planet of high wind.

MAY 20, 2013, MOORE, OKLAHOMA

An EF-5 tornado was created within a super cell thunderstorm that formed over the Plains on May 20, 2013. At 2:45 pm, an EF-0 tornado touched down near Newcastle, Oklahoma, and then tracked northeast, quickly intensifying to EF-4 status. It then switched to an eastward path, and traveled through Moore, an Oklahoma City suburb, attaining winds estimated at 200–210 mph.

Twenty-four people, including ten children, died as a result of the storm, and more than 350 were injured. About 1,500 buildings were severely damaged or destroyed, with another 4,000 receiving varying levels of damage. An elementary school was destroyed, and a medical facility was heavily damaged. There were more than 61,500 power outages, and many overturned vehicles and trailers, some of them on two interstate highways, parts of which were shut down.

The Moore tornado begins with the March 20, 2013, spring ingress chart (inner wheel of Thunderstorms 4). Note that the ingress chart is identical to the one for Lubbock, Texas, shown at the beginning of this chapter (Thunderstorms 1), except for the horizon and meridian. The Moore ingress chart has approximately the same Midheaven degree and sign as the one for Lubbock, Texas, because the two cities are at about the same longitude. However, the Moore ingress chart has 24 Aquarius on the Ascendant, which appears benign in comparison with the Lubbock ingress chart. The Lubbock ingress and lunar phase charts had nearly the same angles by degree and sign, with the Mercury-Neptune conjunction aspecting all four of them. In the Moore chart, none of the four angles was the same, and Mercury-Neptune aspected only the ingress meridian.

Timing is the reason the EF-5 tornado occurred in Moore instead of Lubbock. There were few planetary contacts to the two tornado signatures in the Lubbock ingress chart—Mercury-Neptune and Mars-Uranus—and the Lubbock longitude and latitude. The Moore ingress chart, however, had many significant contacts.

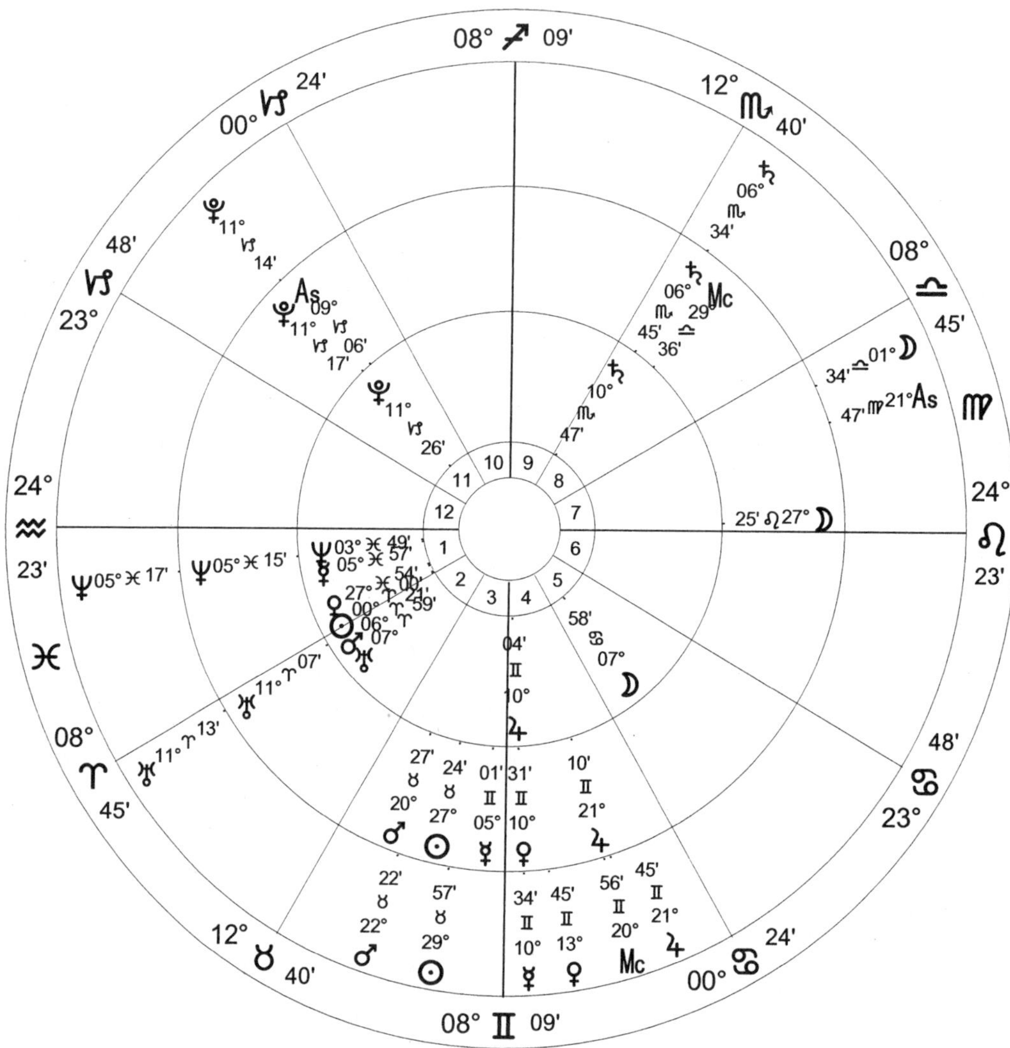

Thunderstorms 4

Inner Wheel=*Spring Ingress / Natal Chart / March 20, 2013, Wed / 6:03 am CDT +5:00*
Moore, OK / 35°N20'22" 097°W29'11" / Geocentric / Tropical / Placidus / Mean Node

Middle Wheel=*Lunar Phase / Natal Chart / May 17, 2013, Fri / 11:36 pm CDT +5:00*
Moore, OK / 35°N20'22" 097°W29'11" / Geocentric / Tropical / Placidus / Mean Node

Outer Wheel=*Tornado / Natal Chart / May 20, 2013, Mon / 2:56 pm CDT +5:00*
Moore, OK / 35°N20'22" 097°W29'11" / Geocentric / Tropical / Placidus / Mean Node

A key factor in the Moore chart is that the Mercury-Neptune conjunction was square the meridian and the Mars-Uranus conjunction was trine the same point. The latitude was involved through Mars-Uranus and Pluto semisquare/sesquisquare the horizon. Uranus and Pluto were square at the ingress, and because Mars was conjunct Uranus, Mars was pulled into the square with Pluto even though it was out of orb with Pluto. On its own, the Uranus-Pluto square is indicative of powerful storms, and Mars increases the intensity. In the Moore chart, Jupiter (high pressure) was conjunct the IC, a much stronger placement than in the Lubbock chart, which drew in Saturn through an inconjunct and the Uranus-Pluto square through a sextile and inconjunct. All of these planetary contacts and others were activated by both the May 17, 2013, lunar phase and the transits.

Before looking at the lunar phase chart and the interaction between it and the ingress chart, there's another significant factor that indicated 2013 would be a strong year for tornadoes: the dry line. This meteorological condition can be found in both the ingress and lunar phase charts. That's not always true, so when you see it, you know there is likely to be a high number of tornadoes.

The dry line is an imaginary line that, when active, runs from southern Kansas south through western Oklahoma and western Texas. It acts as a sort of barrier to keep the warm, dry air from the west from reaching the Plains and allows a strong flow of warm, moist air to move northward from the Gulf. In the summer, the dry line can stretch into Canada. The dry line also represents the approximate location of a low-pressure that is often behind a cold front in severe thunderstorm and tornado weather.

Look at the spring ingress chart and mentally turn it counterclockwise. When you do that, the degrees of the Aquarius Ascendant become smaller, and reach 10 Aquarius, where ingress Saturn will be square the Ascendant. This location, in Colorado at about 105W, marks the dry line. (The dry line generally retreats west at night and moves east during the day; the ingress occurred at 11:36 pm, so the dry line was at or near its farthest west point.) Now look at Neptune in the spring ingress chart. It is square the Midheaven at about 102W and represents a strong flow of warm, moist air rising from the Gulf. This is a perfect setup for a year with many tornadoes, especially when considered along with the two tornado signature conjunctions (Mercury-Neptune and Mars-Uranus). Note also that with Mercury conjunction Neptune, a strong wind would be in force to carry the air north from the Gulf.

The planetary influence that immediately jumps out of the lunar phase chart (middle wheel of Thunderstorms 4) is Pluto (high winds) conjunction the Ascendant and in a nearly exact square with Uranus (cold air); this is an aspect of extreme weather (EF-5 tornado). The lunar phase Sun was semisquare/sesquisquare Uranus-Pluto, an important aspect because the Sun was inconjunct the Midheaven and sesquisquare the Ascendant. Lunar phase Mercury was sextile Uranus and contraparallel the Ascendant (a strong cold front moved through the area). Lunar phase Venus and Mercury conjunct the ingress IC indicated warm and cold air converging in Moore. Saturn was also trine ingress Mercury in water signs (cold, moist air), and both planets aspected the ingress meridian.

There were many connections between the lunar phase and ingress charts. Lunar phase Mars was square the ingress Ascendant and semisquare the ingress Mars-Uranus conjunction. Lunar phase Mercury was approaching a conjunction with the ingress IC and with Jupiter (high winds), the latter of which was exact on the date of the tornado, to activate the Moore longitude; from that point, lunar phase and transiting Mercury were square the ingress Mercury-Neptune conjunction and sextile the ingress Mars-Uranus conjunction. Both tornado signatures were activated by lunar phase Saturn through a trine and an inconjunct.

The important points to remember here are the many lunar phase and transit planetary contacts to the ingress chart longitude and latitude and the ingress tornado signatures. When you see this many, there can be no question that the area will experience not just a tornado but a stronger and highly destructive one.

MAY 6, 2009, MADISON, ALABAMA

Not all tornadoes cause the kind of devastation that was seen in Moore, Oklahoma. More typical are tornadoes much lower on the scale: EF-0, EF-1, and EF-2.

The Madison, Alabama, tornado illustrated here was an EF-2, with winds between 111 and 135 mph. Trees were downed, windows blown out, and roofs damaged. There were no injuries or fatalities.

The March 20, 2009, spring ingress chart (inner wheel of Thunderstorms 5) had a tornado signature: Mercury conjunction Uranus. In Madison, Alabama, this signature was conjunct the Ascendant and square the meridian. So a seasonal forecast would have included a strong possibility for tornadoes in this location.

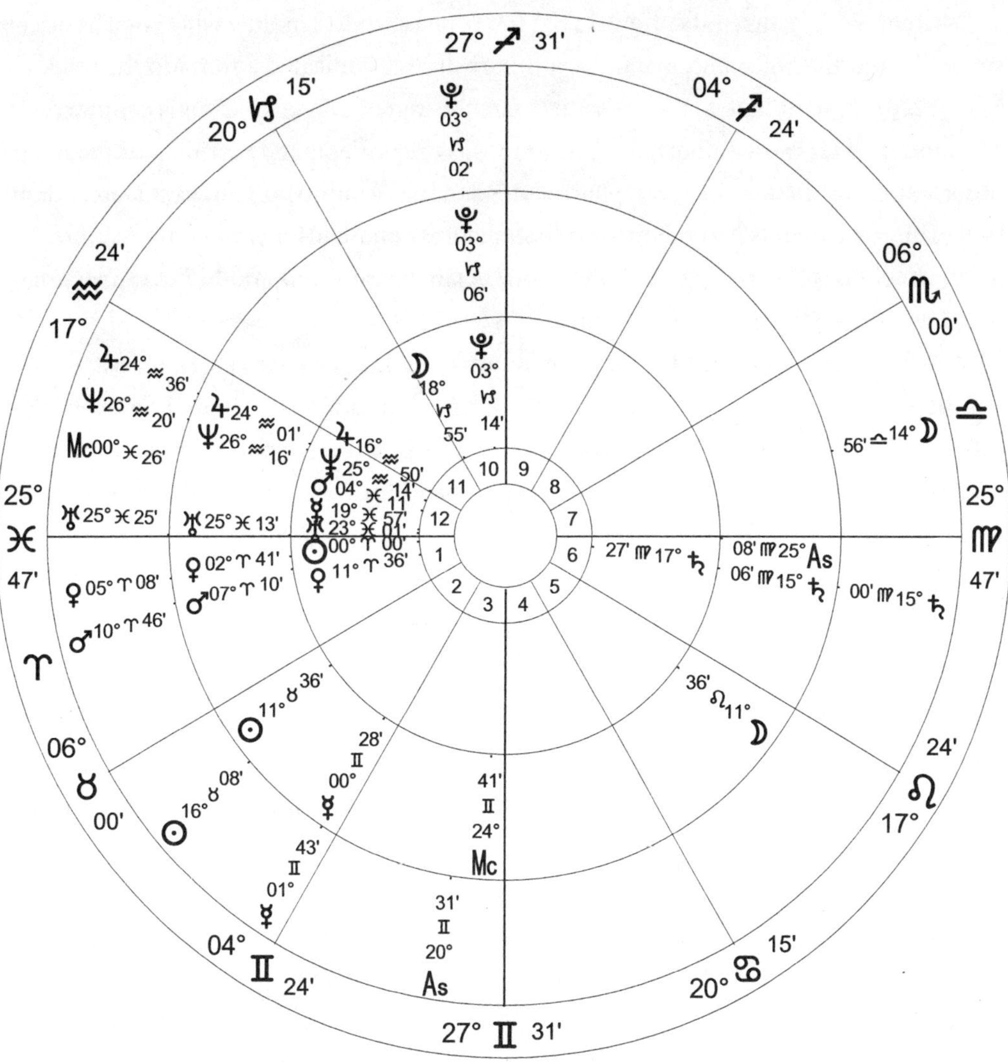

Thunderstorms 5

INNER WHEEL=*Spring Ingress / Natal Chart / March 20, 2009, Fri / 6:43:38 am CDT +5:00*
Madison, AL / 34°N41'57" 086°W44'54" / Geocentric / Tropical / Placidus / Mean Node

MIDDLE WHEEL=*Lunar Phase / Natal Chart / May 1, 2009, Fri / 3:44:13 pm CDT +5:00*
Madison, AL / 34°N41'57" 086°W44'54" / Geocentric / Tropical / Placidus / Mean Node

OUTER WHEEL=*Tornado / Natal Chart / May 6, 2009, Wed / 7:59 am CDT +5:00*
Madison, AL / 34°N41'57" 086°W44'54" / Geocentric / Tropical / Placidus / Mean Node

There were only two indications in the lunar phase chart (middle wheel of Thunderstorms 5) and the lunar phase–ingress comparison that indicated a tornado the week of May 1, 2009. Lunar phase Mars was semisquare the ingress Mercury-Uranus conjunction. The other and far more important indication was the horizon and meridian of the lunar phase chart. Notice that the lunar phase Ascendant was conjunct the ingress Descendant (opposition the ingress Ascendant) and that the lunar phase Midheaven was conjunct the ingress IC (opposition the ingress Midheaven). As in the ingress chart, the lunar phase meridian was square the lunar phase horizon.

This angular connection between the ingress and lunar phase charts almost always indicates a significant weather event. In this case, a tornado was the obvious event because the ingress Mercury-Uranus conjunction was in aspect to both horizons and meridians. Compare this with the many aspects in the Moore chart; more aspects equate to a stronger thunderstorm, with a tornado in many cases.

Tornado Outbreak, February 5–6, 2008

Eighty-seven tornadoes touched down during the course of this outbreak, which lasted more than fifteen hours from the afternoon of February 5 until the early morning of February 6. The majority of the tornadoes occurred in Tennessee, Mississippi, Alabama, Kentucky, and Arkansas, with one or two in each of the states of Louisiana, Indiana, Illinois, Missouri, and Texas. All but ten tornadoes were rated from EF-0 to EF-2, five were rated EF-3, and five rated EF-4. Fifty-six people died.

Winter tornadoes are unusual because even southern states usually lack the warm air necessary for a super cell to develop. However, there was an almost never-ending supply of warm, moist air, along with strong tornado signatures in the ingress and lunar phase charts, and with two stationing planets the week of the outbreak. Temperatures were unseasonably warm in the 70s and 80s.

The charts for Castalian Springs, Tennessee, are representative of the December 22, 2007, winter ingress and January 29, 2008, lunar phase charts (inner and middle wheels of Thunderstorms 6) for the region that experienced the highest number of tornadoes. Degrees on the horizons and meridians were similar for the states involved.

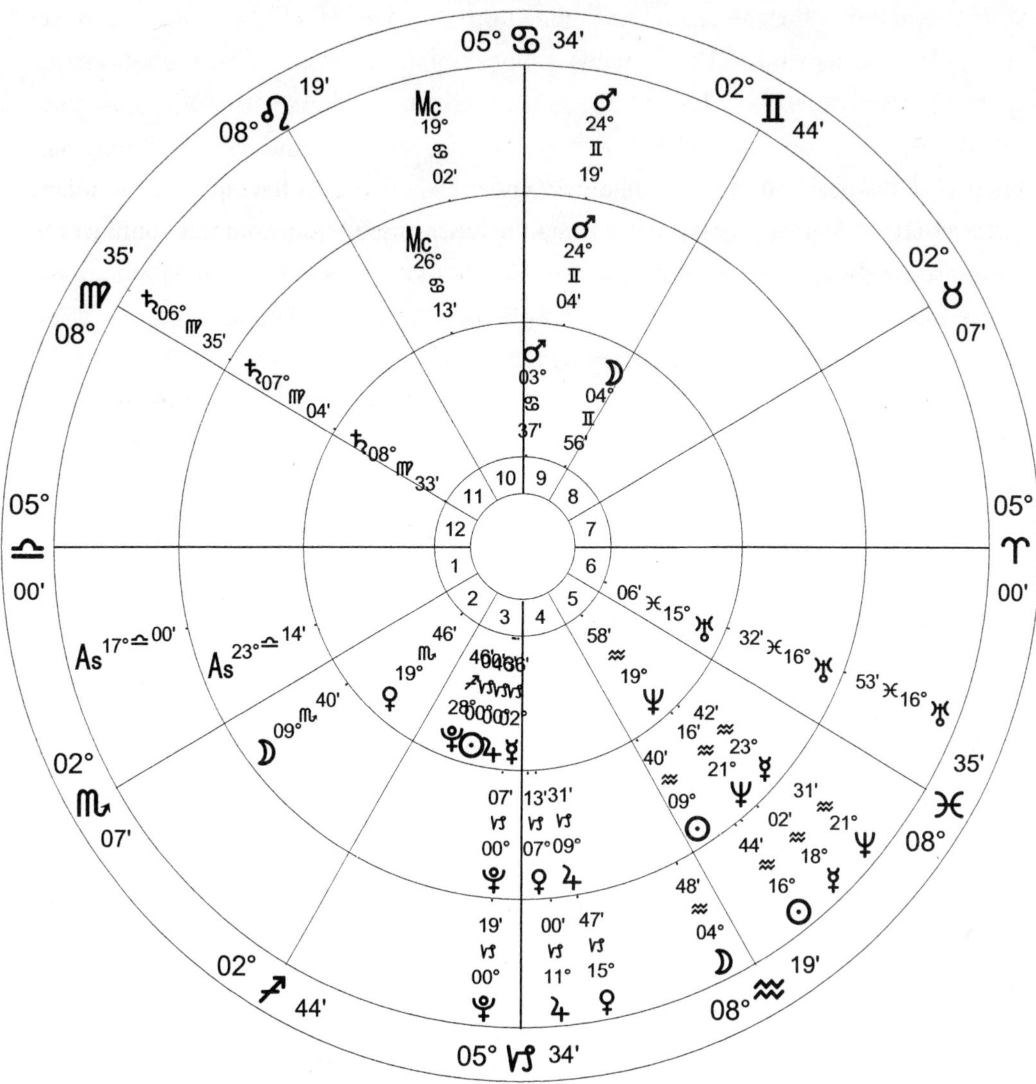

Thunderstorms 6

Inner Wheel=*Winter Ingress / Natal Chart / December 22, 2007, Sat / 0:07:49 am CST +6:00*
Castalian Springs, TN / 36°N23'37" 086°W18'29" / Geocentric / Tropical / Placidus / Mean Node

Middle Wheel=*Lunar Phase / Natal Chart / January 29, 2008, Tue / 11:02:53 pm CST +6:00*
Castalian Springs, TN / 36°N23'37" 086°W18'29" / Geocentric / Tropical / Placidus / Mean Node

Outer Wheel=*Tornado / Natal Chart / February 5, 2008, Tue / 10:05 pm CST +6:00*
Castalian Springs, TN / 36°N23'37" 086°W18'29" / Geocentric / Tropical / Placidus / Mean Node

In the ingress chart, Mercury was opposition Mars (tornado). Also drawn in were the Sun and Jupiter conjunction Mercury (opposition Mars); Mars opposition Jupiter is a thunderstorm aspect, and Jupiter conjunction Mercury indicates high winds, as does out-of-sign Pluto conjunction all three planets. Mercury represents the squall line that was part of the outbreak. This configuration of five planets aspected both the meridian and the horizon. Warm, moist air is indicated by Venus square Neptune, with both planets semisquare/sesquisquare the IC/Midheaven.

The lunar phase chart had three very telling factors: Mercury stationing retrograde, Mars stationing direct, and retrograde Mercury conjunction Neptune (tornado). Stationary planets almost always indicate a major weather event. The Mercury-Neptune conjunction was trine Mars, and all three planets were trine the lunar phase Ascendant and semisextile the meridian.

As strong as all these indicators were of a strong regional super cell with tornado potential, it was the connections between the lunar phase and ingress charts, and transits, that reflect the magnitude of the event.

Pluto had advanced to an exact conjunction with the ingress Sun (high winds), lunar phase Saturn was sesquisquare ingress Mercury (low pressure and wind), and stationary retrograde Mercury conjunction ingress and lunar phase Neptune (tornado) was trine lunar phase stationary direct Mars.

Some areas experienced heavy rain and subsequent flooding. This is indicated by Neptune trine the lunar phase Ascendant and inconjunct the lunar phase Midheaven, lunar phase Venus conjunct the ingress IC and square the ingress horizon, and stationary direct Mars trine Neptune.

Tri-State Tornado, March 18, 1925

The most destructive single tornado to hit the United States in recorded history was the Tri-State Tornado, a name aptly given because it traveled through Missouri, Illinois, and Indiana. Its path of 219 miles ranged from one-quarter to one-half mile wide, with periodic expansion to one mile. In the three and a half hours it moved across three states at an average speed of 67 mph, it killed 689 people, injured 3,000, and destroyed 15,000 homes. It was located close to the point where a warm front and a dryline intersected, both of which then moved eastward, maintaining a favorable super cell environment that sustained the tornado's energy. It is noted for having the longest continuous track on the ground and for being the third fastest tornado in ground speed.

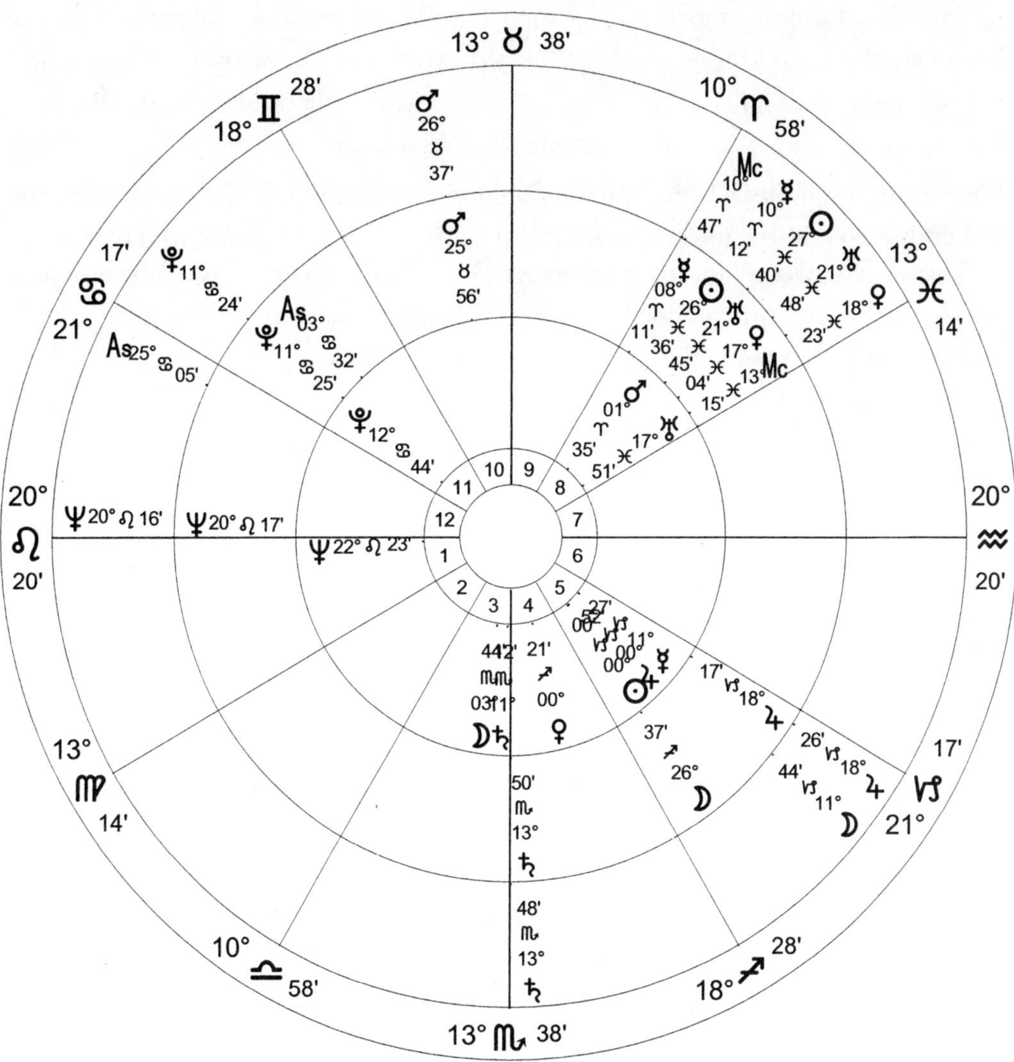

Thunderstorms 7

INNER WHEEL=*Winter Ingress / Natal Chart / December 21, 1924, Sun / 8:46 pm CST +6:00
Annapolis, MO / 37°N21'37" 090°W41'51" / Geocentric / Tropical / Placidus / Mean Node*

MIDDLE WHEEL=*Lunar Phase / Natal Chart / March 17, 1925, Tue / 11:22 am CST +6:00
Annapolis, MO / 37°N21'37" 090°W41'51" / Geocentric / Tropical / Placidus / Mean Node*

OUTER WHEEL=*Tornado Forms / Natal Chart / March 18, 1925, Wed / 1:00 pm CST +6:00
Redford, MO / 37°N19'12" 090°W53'54" / Geocentric / Tropical / Placidus / Mean Node*

The tornado began at Redford in southeastern Missouri about 1:00 pm and moved from southwest to northeast, the direction of travel of most tornadoes. It then hit Annapolis, Missouri, at 1:10 pm before moving on to Biehle in the same state. At 2:20 pm, it crossed the Mississippi River about seventy-five miles southeast of St. Louis, continuing into Illinois and striking Gorham, Murphysboro, DeSoto, and West Frankfort. Griffin and Princeton, Indiana, were its final victims.

The aspects involved in this massive tornado are seemingly endless. Thus, although only the highlights are mentioned here, these charts are excellent ones for study because they show all the necessary thunderstorm and tornado factors.

The December 21, 1924, winter ingress chart for Annapolis, Missouri (inner wheel of Thunderstorms 7), has two indicators for excessively wet and stormy weather, and tornado activity: Neptune conjunct the Ascendant (representing potential for a flow of warm, moist air—the warm front) and Saturn conjunct the IC (this aspect also indicates potential for a dryline to develop in the area). The telltale aspect for tornado potential is Mercury retrograde in nearly exact opposition to Pluto, indicative of very high winds, and which was sextile/trine the meridian. The lunar eclipse of February 8, 1924, at 19 Leo was conjunct the horizon.

The March 17, 1924, lunar phase chart (middle wheel of Thunderstorms 7) has a multitude of aspects that indicate a tornado like no one had ever seen before. In addition to Mercury forming a sesquisquare aspect to Mars, which in itself is indicative of thunderstorms and spurts of whipping winds, it also formed a parallel with Uranus and a sesquisquare with Neptune, two indicators of tornado. Mercury in aspect to both Mars and Uranus promised high winds. Additional indicators of high winds were Mars semisquare and parallel Pluto. And by the lunar phase, Uranus was inconjunct Neptune, an aspect of unusual weather. In the lunar phase chart, Venus was opposite and contraparallel the IC, Saturn and Pluto were sextile/trine the meridian, and Neptune was semisquare/sesquisquare the horizon.

Lunar phase Saturn was semisquare the Sun-Jupiter conjunction in the ingress chart, an important contact because the conjunction was square ingress Mars (wind). Lunar phase Venus and Jupiter aspected ingress Uranus by conjunction and sextile, respectively, and lunar phase Mercury was sesquisquare ingress Neptune, all indicators of potential tornado activity. Another tornado indicator was lunar phase Mars square ingress Neptune. The ingress Sun, Mercury, and Jupiter were contraparallel the lunar phase Ascendant, and ingress Uranus was contraparallel the lunar phase Midheaven. Lunar phase Saturn (low pressure) was contraparallel the ingress Ascendant.

On March 18 (outer wheel of Thunderstorms 7), the potential shown in the ingress and lunar phase charts was realized when both were triggered. Of major significance was transiting Mercury at 10 Aries performing double duty as it formed a square to lunar phase Pluto and the Mercury-Pluto opposition in the ingress chart, which was sextile/trine the lunar phase meridian. To add even more fuel, transiting Pluto at 11 Cancer had retrograded to exactly oppose ingress Mercury.

Transiting Jupiter was parallel ingress Mercury, expanding the latter planet's windy tendencies. Transiting Venus at 18 Pisces was conjunct and parallel ingress Uranus, bringing warm, moist air together with cold, dry air and activating the all-important Uranus influence necessary for the formation of a tornado. The transiting Moon triggered critical planets and points by parallel: ingress Pluto, and lunar phase Mars and Pluto.

Shortly before noon on March 18, the transiting Moon in Capricorn was sesquisquare lunar phase Mars, and at 12:06 pm it was sextile ingress Saturn. About 12:30 pm, two triggering aspects occurred: the transiting Moon was conjunct ingress Mercury and opposition lunar phase Pluto. The tornado began to form at that time, touching down in Redford thirty minutes later. When the tornado hit, the transiting Moon was sesquisquare transiting Mars at 26 Taurus, both of which were in hard aspect to the ingress Ascendant and IC.

The intense low pressure, as well as moisture and a dryline, necessary for a super cell thunderstorm and the creation of a tornado of this magnitude is indicated by two aspects: transiting Neptune at 20 Leo was conjunct within four minutes the ingress Ascendant, having retrograded to that point in the several months since the ingress, and transiting Saturn at 13 Scorpio had moved forward to form a conjunction with the ingress IC within ten minutes of exactitude.

The meridians in the ingress and lunar phase charts were sextile/trine each other, and the horizons were sesquisquare/semisquare, and the Sun and Moon in the lunar phase chart were sesquisquare/semisquare the ingress meridian.

Of special interest in the study of the Tri-State Tornado are the horizons and meridians of some of the other towns it visited:

Location	Horizon	Meridian
Redford, Missouri (ingress)	20 Leo/Aquarius 34	13 Scorpio/Taurus 57
Redford, Missouri (lunar phase)	3 Cancer/Capricorn 20	13 Virgo/Pisces 02
Annapolis, Missouri (ingress)	20 Leo/Aquarius 21	13 Scorpio/Taurus 39
Annapolis, Missouri (lunar phase)	3 Cancer/Capricorn 33	13 Virgo/Pisces 15
DeSoto, Illinois (ingress)	22 Leo/Aquarius 05	15 Scorpio/Taurus 37
DeSoto, Illinois (lunar phase)	5 Cancer/Capricorn 09	14 Virgo/Pisces 50
Princeton, Indiana (ingress)	23 Leo/Aquarius 35	17 Scorpio/Taurus 17
Princeton, Indiana (lunar phase)	6 Cancer/Capricorn 58	16 Virgo/Pisces 38

The southwest to northeast path of the tornado can be seen through the changing degrees of the horizon and meridian. Of particular importance is the effect of Neptune's (counterclockwise movement) conjunction to the ingress Ascendant in all the charts. The tornado died out after hitting Princeton, where Neptune in the ingress chart was just above the horizon. This indicates that the storm changed direction just enough in its eastward movement so that Neptune was no longer below the horizon of the cities it passed through. A similar pattern is apparent in Saturn's conjunction with the IC in various locations.

XENIA, OHIO, APRIL 3, 1974

Tornado sirens sounded across eleven states on two consecutive days in April 1974, warning residents of the potential for major destruction. On April 3 and 4, 1974, there were 127 tornadoes, known as the Super Outbreak. Six of the tornadoes had winds faster than 261 mph, categorizing them as Category 5 storms on the Enhanced Fujita scale. More than 300 people were killed and 6,142 were injured. Total damage was more than $600 million.

The town hardest hit was Xenia, Ohio, where the half-mile-wide tornado destroyed homes, businesses, and the high school. It even lifted freight cars and tossed them across streets. Thirty-four people were killed.

Key planets of this tornado were Pluto, Jupiter, and Mercury. On April 3, 1974, the timing was right, each planet was aspected, and severe weather arrived.

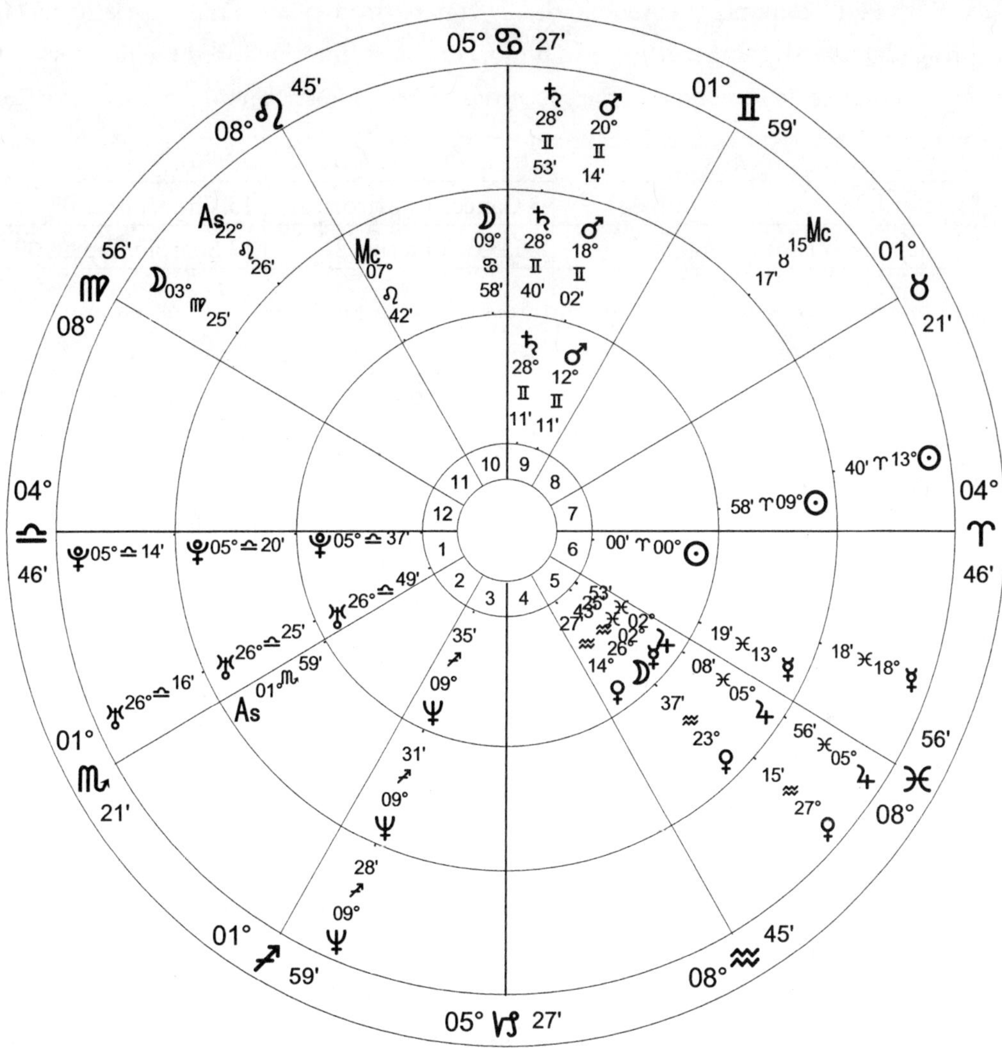

Thunderstorms 8

INNER WHEEL=*Spring Ingress / Natal Chart / March 20, 1974, Wed / 8:07 pm EDT +4:00*
Xenia, OH / 39°N41'05" 083°W55'47" / Geocentric / Tropical / Placidus / Mean Node

MIDDLE WHEEL=*Lunar Phase / Natal Chart / March 30, 1974, Sat / 9:44 pm EDT +4:00*
Xenia, OH / 39°N41'05" 083°W55'47" / Geocentric / Tropical / Placidus / Mean Node

OUTER WHEEL=*Tornado / Natal Chart / April 3, 1974, Wed / 3:40 pm EDT +4:00*
Xenia, OH / 39°N41'05" 083°W55'47" / Geocentric / Tropical / Placidus / Mean Node

With Pluto conjunct the Ascendant and square the meridian, the March 20, 1974, spring ingress chart (inner wheel of Thunderstorms 8) foreshadowed the disaster that would strike Xenia sometime in the succeeding three months. The December 24, 1973, solar eclipse at 3 Capricorn was conjunct the ingress IC and square the ingress horizon.

Uranus and Neptune, either or both of which are active when a tornado strikes, were both aspected by Mars, the first in a sesquisquare and the second in an opposition. Neptune sextile Pluto enhanced the Mars-Neptune aspect, adding moist air and additional counterclockwise wind movement and the high wind promised by Pluto conjunct the Ascendant and square the meridian. More wind is reflected in the Mercury-Pluto inconjunct, a configuration that was transformed into one of excessive wind because of Mercury's conjunction and parallel with Jupiter. The Uranus-Neptune semisquare set the stage for unusual weather.

The March 30, 1974, lunar phase chart (middle wheel of Thunderstorms 8) activated the Uranus and Neptune aspects in the ingress chart. The Sun was trine Neptune, bringing warm, moist air. Mercury sesquisquare Uranus and square Neptune, tornado aspects on their own, activated the Mars aspects to those planets in the ingress chart. Jupiter expanded the same aspects with its parallel to Uranus, delivered an intense cold front, and provided high wind. The long-term Uranus-Neptune semisquare was still in aspect.

The ingress aspects were further activated by lunar phase Mercury square Mars and sesquisquare Uranus. Lunar phase Saturn was sesquisquare ingress Venus, an aspect of heavy precipitation. Ingress Mercury was parallel the lunar phase Ascendant.

On April 3 (outer wheel of Thunderstorms 8), the transiting Sun at 13 Aries triggered ingress Mars, Venus, Mercury, and Jupiter aspects. The Jupiter-Pluto inconjunct was exact that day. When the tornado hit at 3:40 pm, the transiting Moon in Virgo was separating from an opposition to the ingress Mercury-Jupiter conjunction, having activated the inconjunct those planets made to Pluto (conjunct the Ascendant) in the ingress chart. The transiting Sun was also semisquare lunar phase Jupiter, again setting off that aspect, and transiting Mercury at 18 Pisces was square lunar phase Mars.

Note also the many aspects that day involving transiting Venus at 27 Aquarius, which increased precipitation and kept the warm, moist air flowing into the eleven-state tornado area. Venus was situated in the fourth house in many of the lunar phase charts of cities hit by tornadoes. Saturn trine Uranus, in combination with Neptune and Saturn in aspect with other planets, represented the low pressure in the region.

Transiting Jupiter was sextile/trine the ingress meridian that day, transiting Mercury was sesquisquare/semisquare the lunar phase horizon, and transiting Mars was sesquisquare/semisquare the lunar phase meridian. Transiting Venus was parallel the lunar phase Ascendant. Ingress chart aspects that showed potential for a tornado roaring through Xenia were Mars and Saturn (major thunderstorm) contraparallel the IC.

CHAPTER 11

Example Forecasts

Although there are quite a few steps involved in preparing a weather forecast for a given location, with practice you can do this fairly quickly. In the beginning, though, it's best to take the time to list all the various factors for both the ingress and lunar phase charts, along with keywords for each. Doing this is a good way to learn, and the notes will be helpful in hindsight as you compare your forecast to the actual weather.

If you're not interested in a specific date, calculate an ingress chart for your (or another) location. Then flip through the ephemeris to spot dates during that season when one or preferably two inner transiting planets will aspect the ingress longitude or both longitude and latitude. Look for hard aspects from Venus for precipitation, or Mercury and Mars for significant wind. Use those dates for your forecast.

A notable weather event is more likely, of course, when an outer planet is involved and in aspect to the ingress longitude or longitude and latitude. If this isn't the case in your location, use Solar Maps to find, for example, a location where ingress Saturn is square the ingress meridian. Then look for inner planet transits that will trigger Saturn, and use those dates for your forecast.

Following are the steps used in creating a weather forecast:

1. Select a forecast date.

2. Calculate a chart for the cardinal ingress immediately preceding the date.

3. Calculate a chart for the lunar phase immediately preceding the date.

4. List all planetary aspects in the ingress chart, including those to the angles. Write keywords for each planetary configuration, and include all possibilities appropriate to the season and location. After studying which planetary configurations contact the angles, write a brief description or keyword list of the overall weather conditions for the season.

5. List all planetary aspects in the lunar phase chart, including those to the angles. Write keywords for each planetary configuration, and include all the possibilities appropriate to the season and location. After studying which planetary configurations contact the angles, write a brief description or use keywords to describe the overall weather conditions for the week.

6. List all planetary aspects between the ingress and lunar phase charts. Write keywords for each planetary configuration, and note which contacts are activating significant planetary configurations in each of the charts.

7. Using the selected forecast date, list all transiting aspects to the ingress and lunar phase charts, including those to the angles. If the forecast period is for the entire lunar phase, list the aspects by date. Write keywords for each planetary configuration, including all possibilities appropriate to the season and location.

8. List all aspects made by transiting planets to each other on the forecast date. Write keywords for each planetary configuration, including all possibilities appropriate to the season and location. Or, if you're not forecasting for a specific date, look for fast-moving transiting planets that will aspect the lunar phase and ingress charts that week. By doing this, you can spot the dates when there will be precipitation, temperature anomalies, storms, etc.

9. Identify eclipses, if any, that aspect the angles of the forecast location and that have occurred or will occur in the preceding or succeeding six to twelve months.

10. Then formulate a forecast for the selected date or week.

Each of the following three forecast examples represent a different weather event: heat, hurricane, and winter storm. All three were selected using the method explained here, beginning with an ingress map and then scanning the ephemeris for dates when transiting planets would aspect the ingress longitude and/or latitude and outer planets.

2019 Summer Heat, Phoenix, Arizona

Overall, summer 2019 in Phoenix will be remembered as a season with excessively high temperatures. Temperatures during a cool summer in Phoenix are about 105 F; normal is about 110 F. In 2019, though, summer temperatures will be generally above average, with more days in the 115 F or higher range and more nights when the thermometer never drops below the high 90s.

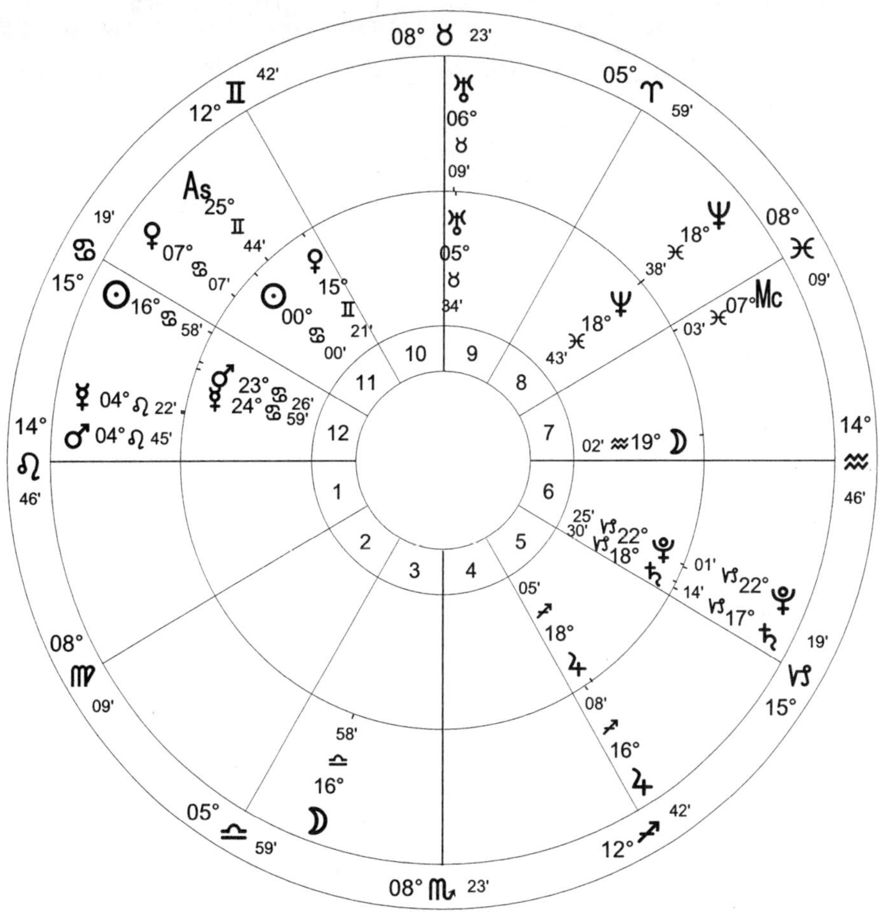

Example Forecasts 1
Inner Wheel=*Summer Ingress / Natal Chart / June 21, 2019, Fri / 8:54:08 am MST +7:00*
Phoenix, AZ / 33°N26'54" 112°W04'24" / Geocentric / Tropical / Placidus / Mean Node

Outer Wheel=*Lunar Phase / Natal Chart / July 9, 2019, Tue / 3:55 am MST +7:00*
Phoenix, AZ / 33°N26'54" 112°W04'24"/ Geocentric / Tropical / Placidus / Mean Node

Ingress Chart

Before looking more closely at all the factors in the summer ingress chart (inner wheel of Example Forecasts 1), note that Uranus is conjunct the Midheaven. This alone indicates the excessive heat that accompanies the high-pressure systems that often influence the region during the summer months. Because this is the ingress chart, these systems will be prevalent throughout the summer.

There are many aspects in the ingress chart, as there are in the lunar phase chart. Although not all of these involve the Phoenix ingress horizon and meridian, it's a good idea to list each one. Doing this will help you as you work with lunar phase charts during the seasons, because weather conditions represented by the ingress aspects can occur when ingress planets aspect the horizon and meridian in a lunar phase chart and vice versa. Remember, weather phenomena will occur where the planets aspect the ingress or lunar phase horizon and/or meridian.

Ingress Planetary Aspects

Five planets in northern declination, five planets in southern declination: average temperatures

Seven planets in high declination, including Mars, Jupiter, and Pluto: high heat

Sun parallel Mars: warmer, storms

Mercury parallel Venus: warmer, partly cloudy, southerly air flow

Mercury conjunct and parallel Mars: windy, hot, dry, thunderstorms

Mercury contraparallel Jupiter: warmer, fair, northerly winds

Mercury contraparallel Saturn: cloudy, low pressure, cooler, precipitation

Mercury opposition Pluto: high winds, hot

Venus parallel Mars: precipitation, heat

Venus opposition and contraparallel Jupiter: cloudy, precipitation, heat

Venus inconjunct Saturn: partly cloudy, cooler

Venus contraparallel Saturn: cloudy, cooler, precipitation, low pressure, humid

Venus square Neptune: abundant precipitation, warmer

Venus contraparallel Pluto: precipitation, windy, warmer

Mars contraparallel Jupiter: heat, thunderstorms

Mars contraparallel Saturn: high wind, storms

Mars opposition and contraparallel Pluto: hot, windy

Jupiter semisextile Saturn: cooler, precipitation

Jupiter parallel Saturn: low pressure, cooler, precipitation

Jupiter sesquisquare Uranus: cooler, storms, windy, precipitation

Jupiter sesquisquare Neptune: abundant precipitation, cloudy, humid

Jupiter parallel Pluto: hot, windy, storms

Uranus semisquare Neptune: cooler, cloudy, humid, storms

Neptune sextile Pluto: hot, humid

Ingress Aspects to Angles

Uranus conjunct/opposition meridian

Sun semisquare/sesquisquare horizon

Venus sextile/trine horizon

Jupiter trine/sextile horizon

Saturn inconjunct/semisextile horizon

Neptune inconjunct/semisextile horizon

On its own, Uranus prominent in the chart indicates a summer with temperatures above average. This is compounded by four other planets that represent heat (Sun, Venus, Jupiter, Neptune) and, by extension, two other planets of heat, Mars and Pluto. Mars is drawn in because it is parallel the Sun and Venus, and Jupiter is parallel Pluto, in addition to other aspects. So even though a specific planet in aspect to another planet does not aspect the horizon and/or meridian, it still has influence because it intensifies the meaning of the planet that does aspect the angle. The inclusion of planetary aspects by extension is best used as a confirmation of what the chart already indicates, because these aspects aren't as strong as those that do contact the angles. Otherwise you could find yourself going around and around in circles, creating a long chain of links!

The only ingress aspect that indicates cooler temperatures is Saturn. Considered along with Taurus/Scorpio on the meridian and the Neptune aspects, this indicates that the area will receive some monsoon storms and cloudy days with higher humidity.

Now let's look at the aspects in the lunar phase chart for July 9 (outer wheel of Example Forecasts 1).

Lunar Phase Planetary Aspects

Sun inconjunct Jupiter: warmer

Sun contraparallel Jupiter: warmer

Sun opposition and contraparallel Saturn: low pressure, cooler, cloudy, abundant precipitation, storms

Sun trine Neptune: warmer, humid

Sun contraparallel Pluto: warmer

Mercury semisextile Venus: warmer, partly cloudy

Mercury conjunct Mars: windy, thunderstorms, hot, dry

Mercury sesquisquare Jupiter: warmer, fair to partly cloudy

Mercury square Uranus: thunderstorms, windy, high pressure

Mercury sesquisquare Neptune: warmer, humid, cloudy, precipitation

Venus semisextile Mars: warmer, fair

Venus sextile Uranus: cooler, fair

Mars sesquisquare Jupiter: hot, thunderstorms

Mars square Uranus: thunderstorms, windy

Mars sesquisquare Neptune: humid, warmer, precipitation

Jupiter semisextile Saturn: cooler, precipitation

Jupiter parallel Saturn: low pressure, cooler, precipitation, storms

Jupiter square Neptune: abundant precipitation, cloudy, humid

Jupiter parallel Pluto: hot, windy, storms

Saturn sextile Neptune: precipitation, cooler

Saturn contraparallel Pluto: storms, windy, cloudy

Uranus semisquare Neptune: cool, cloudy, humid, storms

Neptune sextile Pluto: hot, humid

Lunar Phase Aspects to Angles

Mercury inconjunct/semisextile meridian

Venus trine/sextile meridian

Mars inconjunct/semisextile meridian

Uranus sextile/trine meridian

Venus parallel/contraparallel horizon

Pluto inconjunct/semisextile horizon

The lunar phase chart mirrors what the ingress chart forecast: above-normal temperatures. All of the planets that contact the angles, except Mercury, are associated with heat, and Mercury has a similar characteristic because it's in Leo and in aspect to planets that represent higher temperatures. Note also that Mercury is retrograde, which reinforces the hard aspects. Mercury and its aspects also indicate a high probability for dust storms, which are common in monsoon season and often precede thunderstorms.

Lunar Phase–Ingress Aspects

Aspects involving the outer planets repeat in a comparison of the two charts. This isn't always true and varies according to the time between the ingress and the lunar phase and the movement of the outer planets. In this case, since there will be only about three weeks between the two dates, the aspects are nearly identical. The inner planets, however, will advance far enough to make new aspects between the charts, with some of the planets changing signs. Thus, the aspects listed here are only those that involve the faster-moving planets and aren't present in either the ingress or lunar phase chart (I = ingress, LP = lunar phase). Equally, and sometimes more, important are interchart aspects from planets to the horizon and/or meridian.

LP Sun semisextile I Venus: warmer, fair

LP Mercury semisquare Venus: warmer, some humidity, partly cloudy

LP Mars semisquare I Venus: precipitation, storms

LP Sun semisextile/inconjunct I horizon

LP Mercury square I meridian

LP Venus sextile/trine I meridian

LP Mars square I meridian

LP Jupiter trine/sextile I horizon

LP Saturn inconjunct/semisextile I horizon

LP Neptune inconjunct/semisextile I horizon

LP Pluto inconjunct/semisextile LP horizon

LP meridian sextile/trine I meridian

LP horizon semisquare/sesquisquare I meridian

Here, again, planets of heat are emphasized, along with a lesser emphasis on wind and thunderstorms.

Eclipses

There are three eclipse contacts, one to the ingress chart and two to the lunar phase chart, and only one of these is a hard aspect. Overall, because two of the three eclipse contacts are easy aspects, this is a weaker influence, and although it indicates the potential for a significant weather event, it is unlikely to be on the scale of a hurricane or an EF-5 tornado.

Solar eclipse, 15 Capricorn 25, January 6, 2019, inconjunct/semisextile ingress horizon

Solar eclipse, 10 Cancer 38, July 2, 2019, trine/sextile lunar phase meridian

Lunar eclipse, 24 Capricorn 04, July 16, 2019, inconjunct/semisextile lunar phase horizon, sesquisquare/semisquare lunar phase meridian

Transits

Mars will form an exact square with the ingress meridian on July 15, 2019, the day before the next lunar phase. The high temperature will increase each day and reach the highest for the week on this date. As Mars does this, it will activate all the aspects to the ingress meridian, most of which also indicate above-average temperatures. On the same date, the transiting Moon will be in Capricorn, where it will oppose lunar phase Venus trine/sextile lunar phase meridian, and retrograde Mercury at 2 Leo will aspect the ingress Jupiter-Neptune square (both of those planets aspect the ingress horizon). So in addition to excessively high temperatures and humidity on the 15th, Phoenix could experience a dust storm followed by thunderstorms.

2018 Gulf Hurricane

The National Hurricane Center issues a hurricane probability forecast each year, citing a probable number of named storms and other data. You can do the same using the summer and autumn ingress charts, the two seasons when nearly all hurricanes occur. The first step is, of course, to look for hurricane signatures in the ingress chart and then to see which ingress longitudes and latitudes they aspect.

Autumn 2018 (inner wheel of Example Forecasts 2) has potential for a Gulf hurricane or tropical storm because of these aspects: Mercury inconjunct Uranus, Mars square Uranus, and Mars semisquare Neptune; Mercury square Saturn, although not a hurricane aspect, is associated with low-pressure systems.

Since we're looking for a location with hurricane potential, we'll begin with a map that shows where these hurricane signatures aspect the ingress longitude and latitude.

The map (Example Forecasts 3) shows Mars conjunct/opposition the meridian in the central area of the Florida Panhandle, with Uranus square the meridian in the western area of the Florida Panhandle and to the east of Mobile, Alabama. So this part of the Gulf has hurricane potential.

Just west of Mobile, Neptune is sextile/trine the horizon and also semisquare/sesquisquare the meridian. East of Mobile (within orb), Mercury is trine/sextile the meridian and semisquare/sesquisquare the horizon. Neptune doesn't aspect the horizon or meridian in the Florida Panhandle. Saturn aspects the horizon or meridian both near Mobile and in the Panhandle. But because both Mercury and Neptune aspect Mobile, along with the Mars-Uranus square (within orb), that location is more likely than the Florida Panhandle to experience a hurricane landfall—at least from the perspective of the ingress chart. Both locations, however, will be affected by any hurricane in the area, with severe thunderstorms with tornado potential.

Next, scan an ephemeris for dates when transiting planets have the potential to trigger the hurricane signatures. The lunar phase of October 8 (outer wheel of Example Forecasts 2) has potential because Mercury will advance into Scorpio that week and square the ingress meridian, and because Venus will have turned retrograde at 10 Scorpio on October 5. Not a perfect fit but one with enough potential to calculate the lunar phase chart.

Before looking further into this possibility, we'll backtrack and list the aspects in the ingress and lunar phase charts for Mobile, as well as those between the charts.

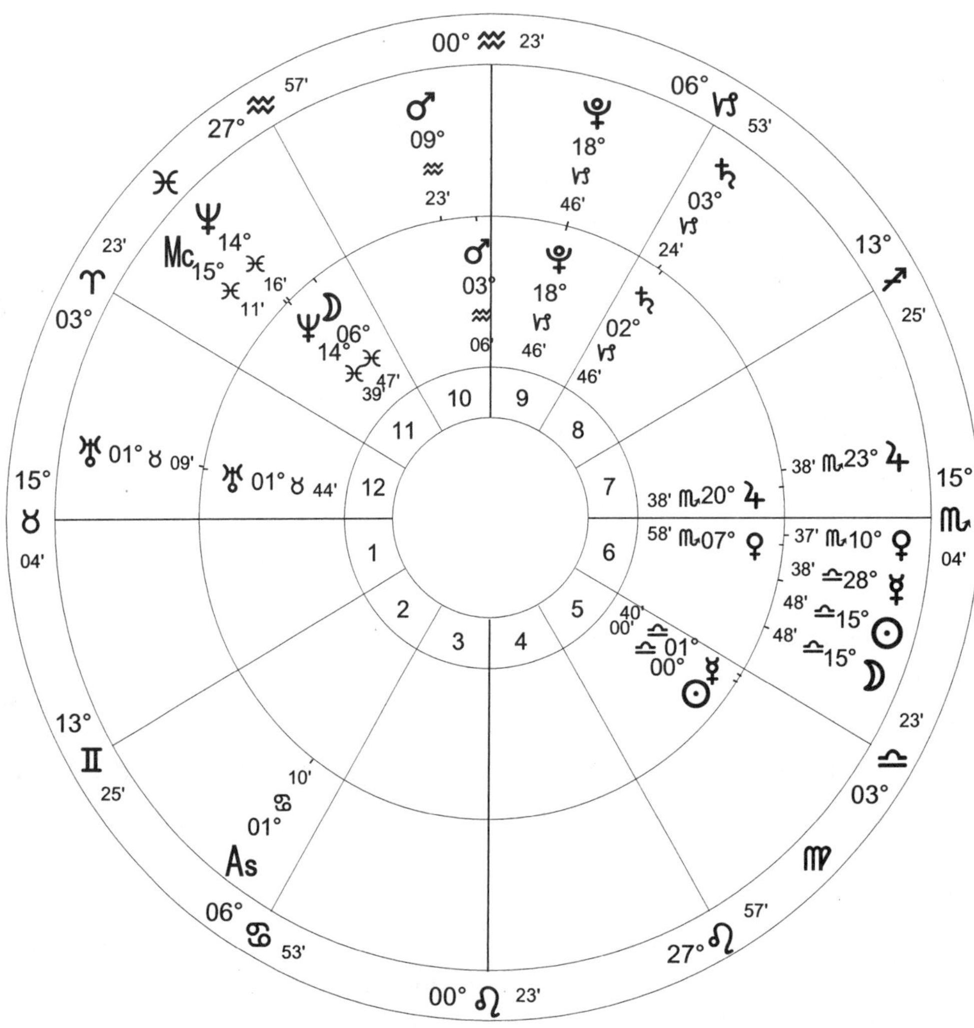

Example Forecasts 2

Inner Wheel=*Autumn Ingress / Natal Chart / September 22, 2018, Sat / 8:55 pm CDT +5:00
Mobile, AL / 30°N41'39" 088°W02'35" / Geocentric / Tropical / Placidus / Mean Node*

Outer Wheel=*Lunar Phase / Natal Chart / October 8, 2018, Mon / 10:46:45 pm CDT +5:00
Mobile, AL / 30°N41'39" 088°W02'35" / Geocentric / Tropical / Placidus / Mean Node*

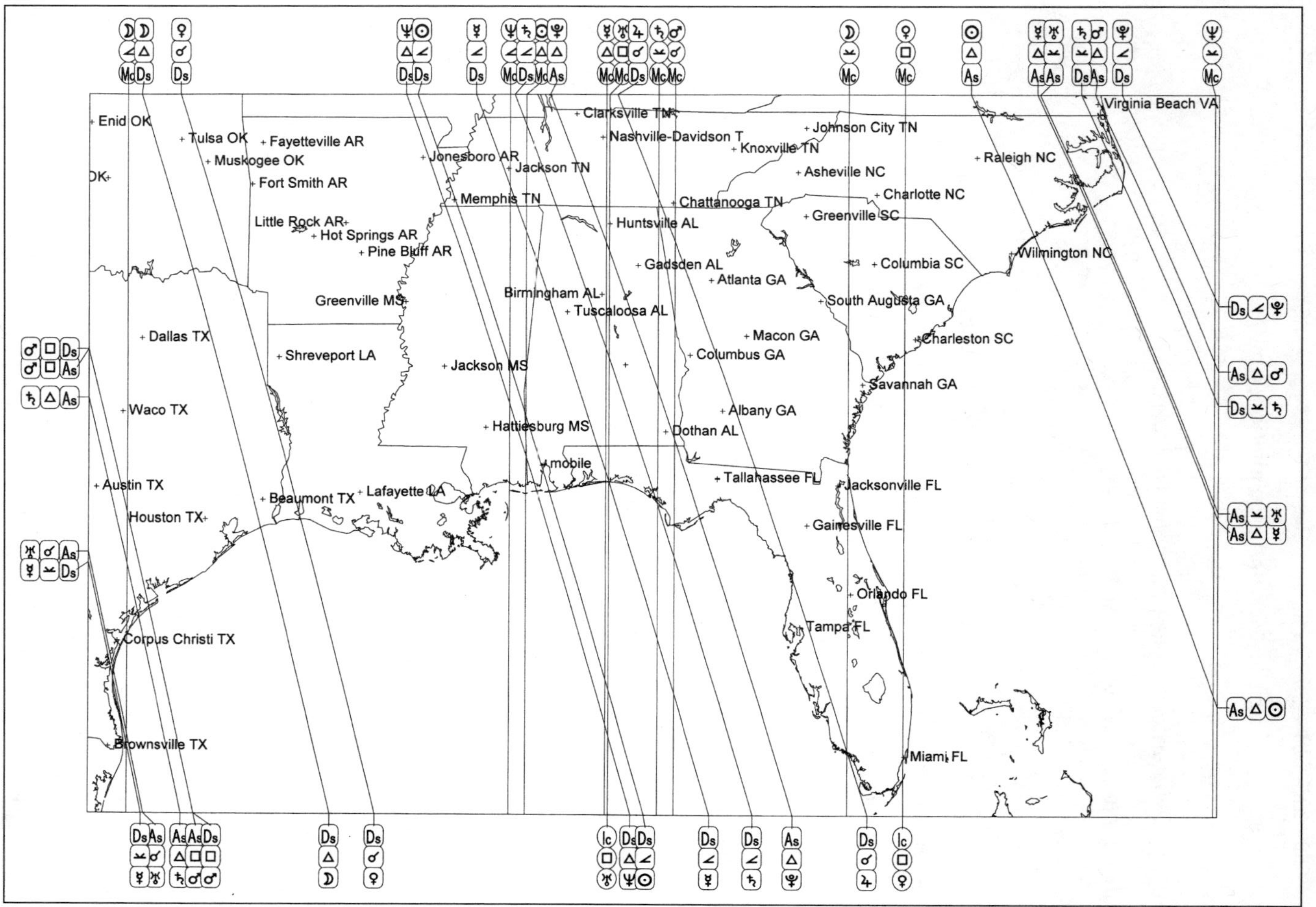

Example Forecasts 3: Solar Map for the 2018 Autumn Ingress

Ingress Planetary Aspects

Sun conjunct and parallel Mercury: precipitation, wind

Sun trine Mars: warm, dry

Sun square Saturn: low pressure, cooler, cloudy, storms

Sun inconjunct Uranus: cooler, fair

Mercury trine Mars: warm, dry

Mercury semisquare Jupiter: warm, fair to partly cloudy

Mercury square Saturn: precipitation, cloudy, low pressure, humid, cooler

Mercury inconjunct Uranus: cloudy, storms, hurricanes, tornadoes

Mars semisextile and parallel Saturn: cloudy, windy, storms

Mars square Uranus: hurricanes, windy, storms, tornadoes

Mars semisquare Neptune: hurricanes, humid, precipitation, tornadoes

Jupiter semisquare Saturn: storms, cloudy, low pressure, cooler

Jupiter sextile Pluto: warm or cool, dry, windy

Saturn trine Uranus: cool, precipitation

Saturn parallel Pluto: storms, wind

Uranus semisquare Neptune: cool, cloudy, humid, storms, wind

Aspects to Ingress Angles

Sun trine/sextile meridian

Sun sesquisquare/semisquare horizon

Mercury trine/sextile meridian

Mercury sesquisquare/semisquare horizon

Venus parallel/contraparallel meridian

Mars conjunct/opposition meridian

Jupiter contraparallel/parallel horizon

Saturn semisextile/inconjunct meridian

Saturn sesquisquare/semisquare horizon

Uranus square meridian

Neptune semisquare/sesquisquare meridian

Neptune sextile/trine horizon

With eight of ten planets aspecting the horizon and meridian at Mobile, the overall trend is for temperatures ranging from average to below and wind, and potential for considerable precipitation associated with a tropical storm or hurricane as well as other severe thunderstorms with tornado potential.

Lunar Phase Planetary Aspects

Sun inconjunct and parallel Neptune: warm, humid

Sun square Pluto: warm, high winds, storms

Mercury opposition and contraparallel Uranus: hurricanes, tornadoes, storms, wind

Mercury sesquisquare Neptune: hurricanes, tornadoes, humid, warm, cloudy, precipitation

Venus square and parallel Mars: precipitation, storms

Venus parallel Saturn: precipitation, cooler, low pressure, humid

Venus trine Neptune: warm, humid, cloudy, precipitation

Venus parallel Pluto: precipitation, wind

Mars parallel Pluto: high wind, storms

Saturn trine Uranus: cool, precipitation

Saturn parallel Pluto: storms, wind

Uranus semisquare Neptune: cool, cloudy, humid, storms, wind

Lunar Phase Aspects to Angles

Sun inconjunct/semisextile and parallel/contraparallel meridian

Mercury sesquisquare/semisquare meridian

Mercury trine/sextile horizon

Saturn opposition/conjunct and contraparallel horizon

Uranus semisquare/sesquisquare meridian

Uranus sextile/trine horizon

Neptune conjunct/opposition meridian

Pluto sextile/trine meridian

The lunar phase chart indicates abundant precipitation, low pressure, and humidity along with high potential for a storm, possibly a hurricane. Temperature is a mix of warm and cool, which likely reflects conditions before (warm) and after (cool) a storm. Whenever you see Neptune in aspect to the meridian, heavy precipitation is nearly guaranteed, and in this case it could be a major storm.

Lunar Phase–Ingress Aspects

Following are the inner planet aspects from the lunar phase chart to the ingress chart (LP = lunar phase, I = ingress):

LP Sun inconjunct/semisextile I horizon

LP Venus conjunction/opposition ingress horizon (approaching aspect)

LP Mercury square ingress meridian (approaching aspect)

LP horizon inconjunct/semisextile I meridian

LP horizon semisquare/sesquisquare I horizon

LP meridian sextile/trine I horizon

LP meridian semisquare/sesquisquare I meridian

The four aspects that link the horizon and the meridian in the two charts offer confirmation that Mobile will be the site of a hurricane or major storm. Note the approaching aspects; these aspects will become exact during the week of the lunar phase.

Eclipses

The eclipses offer more confirmation for stormy conditions in Mobile. Note that two of the contacts are strong hard aspects, both involving the ingress chart and thus indicating a seasonal trend for one or several major weather events.

Lunar eclipse, 4 Aquarius 45, July 27, 2018, conjunct/opposition ingress meridian, inconjunct/semisextile lunar phase horizon, sesquisquare/semisquare lunar phase meridian

Solar eclipse, 18 Leo 42, August 11, 2018, square ingress horizon, inconjunct/semisextile lunar phase meridian, sesquisquare/semisquare lunar phase horizon

Transits

On October 10, 2018, the transiting Scorpio Moon will square the ingress meridian as Mercury, which will also be in Scorpio, will square the ingress meridian. Both of these planets will form a trine/sextile with the lunar phase Ascendant.

Venus and Mars will form a square on the same date at 10 Scorpio-Aquarius, which indicates stormy conditions and heavy precipitation. Although this square is within orb of the ingress horizon, it's a bit wide; however, it's also a stronger aspect because Venus will be retrograde, having made its station only five days earlier.

The many aspects in both charts and between them, along with the eclipses and transits, indicate that conditions will be stormy with abundant precipitation in Mobile. The question is whether it will be a hurricane or tropical storm. Despite the strong presence of Saturn, Uranus, and Neptune, and aspects to these planets, I would forecast a tropical storm. For a hurricane, I would expect to see Mars square Neptune in the ingress chart, rather than a semisquare, as well as stronger hard aspects between these planets at the lunar phase, rather than a Mercury-Neptune sesquisquare. This forecast is also more logical from a meteorology perspective. By early to mid October, the Gulf will not have the very warm sea surface temperatures necessary to support a hurricane.

Winter 2017–2018, Chicago, Illinois

Chicago's location lends itself to weather anomalies and widely varying weather conditions. Its proximity to Lake Michigan can result in warmer temperatures but also higher levels of precipitation. Storms and weather systems, both low and high pressure, move through the area because the jet stream is often near or over Illinois, especially in fall, winter, and spring.

With ingress Neptune square the meridian and retrograde Mercury also aspecting the meridian (inner wheel of Example Forecasts 4), snowfall amounts could set a new record during winter 2017–2018. The winters of 1977–1978 and 1978–1979, two of the snowiest on record, also had a strong Neptune influence: square the meridian in 1977–1978 and conjunct/opposition the meridian and square the horizon in 1978–1979.

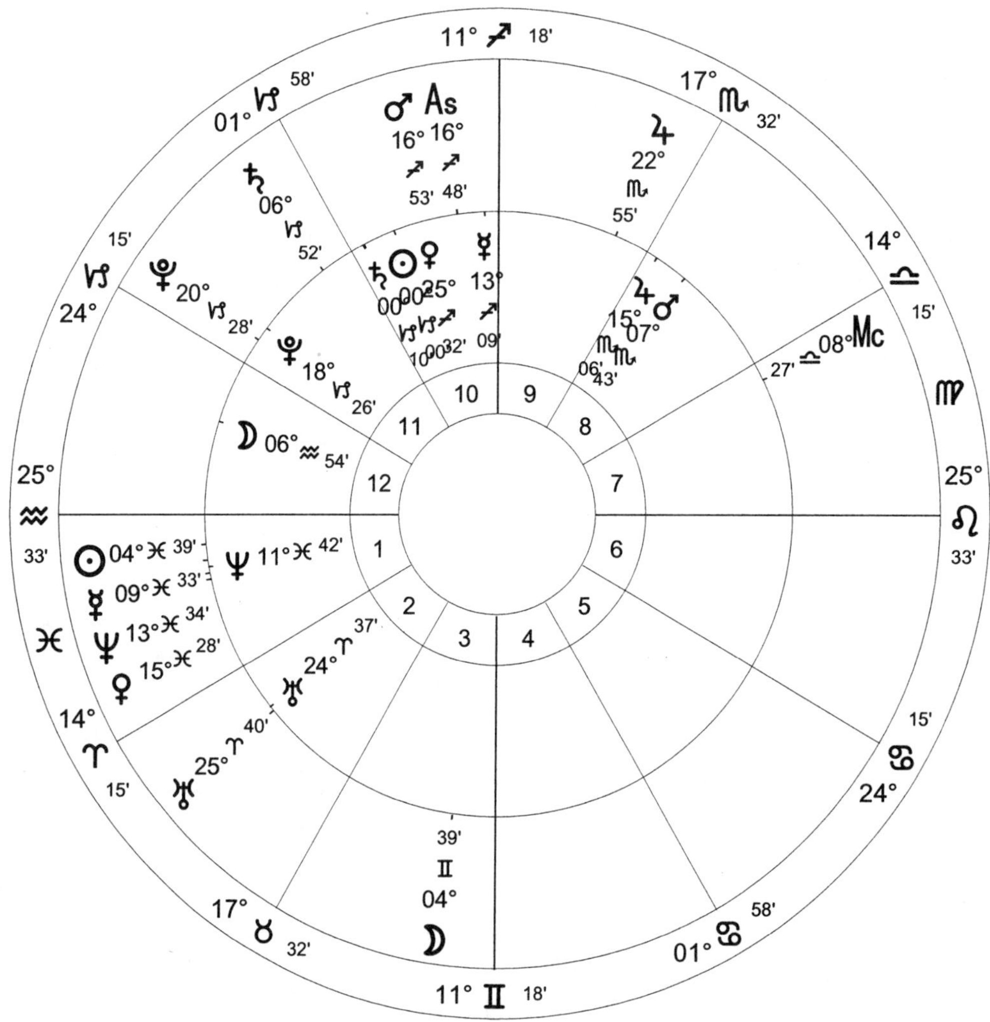

Example Forecasts 4
Inner Wheel=*Winter Ingress / Natal Chart / December 21, 2017, Thu / 10:27:53 am CST +6:00*
Chicago, IL / 41°N51' 087°W39' / Geocentric / Tropical / Placidus / Mean Node

Outer Wheel=*Lunar Phase / Natal Chart / February 23, 2018, Fri / 2:09 am CST +6:00*
Chicago, IL / 41°N51' 087°W39' / Geocentric / Tropical / Placidus / Mean Node

Ingress Planetary Aspects

Sun parallel Venus: cool, heavy precipitation

Sun semisquare Jupiter: warm, windy, thunderstorms

Sun conjunct and parallel Saturn: excessive precipitation, low pressure, cool, cloudy, storms

Mercury retrograde semisextile Jupiter: seasonal to cool, partly cloudy

Mercury retrograde sesquisquare Uranus: storms, tornadoes, high pressure, cloudy

Mercury retrograde square Neptune: cloudy, abundant precipitation, cool, humid/damp

Venus semisquare Mars: precipitation, storms

Venus parallel Saturn: cloudy, heavy precipitation, sleet, snow

Venus trine Uranus: fair, cold

Mars trine Neptune: warm, humid/damp

Jupiter semisquare Saturn: storms, cloudy, low pressure, below-average temperatures

Jupiter trine Neptune: above-average temperatures, fair to partly cloudy, thaw

Jupiter sextile Pluto: above-average temperatures, dry

Saturn parallel Pluto: storms, windy

Uranus semisquare and contraparallel Neptune: below-average temperatures, cloudy, humid/damp, storms, windy

Ingress Aspects to Angles

Eight planets aspect the ingress angles. This seldom occurs, and when it does, it's an indication that the location will see not one or two or a few notable weather events, but many.

Mercury conjunct/opposition meridian

Venus sextile/trine horizon

Mars semisextile/inconjunct meridian

Mars parallel/contraparallel horizon

Jupiter semisextile/inconjunct meridian

Saturn parallel/contraparallel meridian

Uranus sesquisquare/semisquare meridian

Uranus sextile/trine horizon

Neptune square meridian

Pluto parallel/contraparallel meridian

The ingress chart indicates a winter with abundant precipitation. Temperatures will vary from below to above normal, with the latter associated with winter storms. However, with so many indicators for warmer temperatures, warm, moist air is likely to move northward from the Gulf (Neptune square the meridian) and mix with colder jet stream air (Mercury-Uranus) to produce record-setting snow levels. A blizzard is possible with Mercury retrograde in the ingress chart, and the odds increase because Mercury, almost at a standstill, will station direct two days later.

Lunar Phase Aspects

A scan of the ephemeris reveals that the February 23, 2018, lunar phase (outer wheel of Example Forecasts 4) has potential to trigger a major storm with abundant precipitation. Three planets—Mercury, Venus, and Mars—will be in aspect to Neptune either at the ingress or during that week. Note also that the these three planets are in the third house: weather approaching from the west.

Sun parallel Mercury: precipitation

Sun sextile Saturn: cooler, partly cloudy

Sun contraparallel Uranus: fair, sudden cold wave

Sun semisquare Pluto: warmer, precipitation

Mercury sextile Saturn: partly cloudy, cooler

Mercury semisquare and contraparallel Uranus: storms, high wind

Venus square Mars: precipitation, storms

Venus conjunct and parallel Neptune: warmer, abundant precipitation

Mars parallel Saturn: storms, high wind

Mars square Neptune: humid/damp, precipitation, warm, moist air flow

Mars semisextile and parallel Pluto: warm, windy, storms

Jupiter semisquare Saturn: storms, cloudy, low pressure, below-average temperatures

Jupiter trine Neptune: above-average temperatures, fair to partly cloudy, thaw

Jupiter sextile Pluto: above-average temperatures, dry

Saturn parallel Pluto: storms, windy

Uranus semisquare and contraparallel Neptune: below-average temperatures, cloudy, humid/damp, storms, windy

Lunar Phase Aspects to Angles

As in the ingress chart, eight planets aspect the angles at the lunar phase, indicating potential for a major weather event. Of these, precipitation planets Venus and Neptune are particularly strong, increasing the odds for abundant snowfall. Saturn (also strong with its square to the meridian) and Neptune are planets associated with low-pressure, major storm systems.

Sun inconjunct/semisextile meridian

Mercury inconjunct/semisextile meridian

Venus square horizon

Mars conjunct and parallel horizon

Jupiter semisquare/sesquisquare meridian

Saturn square meridian

Saturn parallel/contraparallel horizon

Neptune square horizon

Pluto semisextile/inconjunct horizon

The lunar phase chart mirrors the ingress chart forecast: heavy precipitation, stormy conditions, and a southerly flow of warm, moist air from the Gulf to increase snowfall totals. There is also a mix of cold and warm indicators, as well as potential for high winds because of Mars conjunct the Ascendant.

Lunar Phase–Ingress Aspects

Here are the lunar phase–ingress aspects (LP = lunar phase, I = ingress), other than those involving only the outer planets. Note especially that the lunar phase chart activates the ingress retrograde Mercury-Neptune square of heavy precipitation, and that Mercury advanced about 90 degrees since the ingress to form a conjunction at the lunar phase.

Although lunar phase Mars is separating from a conjunction/opposition to the ingress meridian and Mercury, I would consider this a valid aspect for these reasons: Mars is strongly placed in the lunar phase chart; and Mars, a planet of wind and warm air, will form a square with Neptune and Venus (also warm planets) prior to the lunar phase, representing the warm, moist air flow from the Gulf prior to the storm. Temperatures will drop under the February 23 lunar phase as Mercury interacts with Mercury and Uranus.

LP Sun trine I Mars

LP Sun semisquare I Uranus (approaching)

LP Mercury conjunct I Neptune

LP Mercury square I Mercury

LP Mercury semisquare I Uranus

LP Venus square I Mercury

LP Venus trine I Jupiter

LP Venus sextile I Pluto

LP Mars conjunct I Mercury (separating)

LP Mars semisextile I Jupiter

LP Jupiter semisextile I Venus

LP meridian sextile I meridian

LP Mercury square I meridian

LP Neptune square I meridian

LP Venus square I meridian

LP Uranus sextile/trine I horizon

LP Jupiter square I horizon

Eclipses

Lunar eclipse, 15 Aquarius 25, August 7, 2017, sextile/trine ingress meridian and lunar phase horizon

Solar eclipse, 28 Leo 53, August 21, 2017, conjunct/opposition ingress horizon

Lunar eclipse, 11 Leo 37, January 31, 2018, trine/sextile ingress meridian, sextile/trine lunar phase meridian

Solar eclipse, 27 Aquarius 08, February 15, 2018, conjunct/opposition ingress horizon

Temperatures will be mostly seasonal in Chicago during the winter but cold when major winter storms (cyclones) develop in the region, bringing considerable precipitation. Warmer days and weeks could lead to fast snowmelt, which in turn will make the area prone to flooding. Days with temperatures well above average are likely to precede the most major of the storms.

CHAPTER 12

Climate Change

What, if anything, can astrometeorology contribute to the ongoing debate about climate change? Something, certainly, but probably not more than atmospheric science and history can. Astrometeorology can, however, provide information regarding historical records for particularly cold, dry, wet, or warm years.

The European Great Famine of 1315–1317 had lasting consequences, including a population decline that is estimated at between 10 and 25 percent. Thousands starved because crop yields were small to nonexistent as a result of cold, wet weather. The famine occurred during a twenty-year period (1310–1330) when winters were severe and summers were cold and wet. Some scientists and historians identify 1300 as the end of the Medieval Warm Period and the onset of the Little Ice Age. Although there is not complete agreement about the dates of the Little Ice Age, many scientists say it lasted until about the mid-1800s, with a warmer period in the 1500s, followed by the coldest temperatures of the Little Ice Age.

From 1315 to 1317, all four of the outermost planets were in southern declination and Saturn was sextile Uranus; Jupiter was square Neptune from the fall of 1315 until the summer of 1316. Jupiter was square Neptune from the autumn of 1315 through the spring of 1316. In the summer of 1315, Jupiter was contraparallel Saturn and Neptune, and in the autumn of 1315, Saturn was parallel Neptune. All of these aspects confirm the information in the historical records.

The winter of January 1186–1187 was reportedly a very warm winter, with trees blooming in Strasbourg, France, in January. This is unlikely. No doubt there were some unseasonably warm days and weeks, because Mars was conjunct the IC at the ingress. But Jupiter conjunct Saturn, which was square Mercury trine Uranus, indicate otherwise. Three of the five outermost planets were in southern declination.

Records indicate that Hungary was cold and snowy during the winter of 1342, a seasonal trend confirmed by Mars conjunct Pluto square Saturn. New York Harbor froze in the winter of 1780, allowing people to walk across it from Manhattan to Staten island; four of the five outermost planets were in southern declination and Saturn was sextile Pluto. During the winter of 1794–1795, the Dutch fleet was stranded in Den Helder harbor, frozen in ice; Saturn was square Pluto in cold Aquarius and Uranus, which was square Pluto; three of the five outermost planets were in southern declination.

And then there is the well-documented severe weather of the 1930s in the United States, which included very hot and very cold temperatures, flooding, and drought—and which is being repeated in the early decades of the twenty-first century.

All of these examples and many of those in this book can be characterized as extreme weather, a phenomenon that is definitely not something new to this century, as some proponents of climate change would have the public believe.

Weather, like everything else in life, is cyclical. When extreme weather events are few, we consider the weather to be "normal." When extreme weather events are many, especially during a series of years, we tend to think that the climate is undergoing a permanent change.

The Uranus-Pluto square of the 1930s returned in the late 2000s for a lengthy stay; extreme weather repeats. Notice how prominent Uranus and Pluto are in the examples just cited; extreme weather repeats. If people had been more aware of the environment in the 1930s, talk of climate change would have been prevalent. Same with the Medieval Warm Period and the onset of the Little Ice Age. During the twentieth century, global temperatures have risen between 1 F and 1.4 F, which is comparable to the Medieval Warm Period.

Climate change became a hot topic during the time that Pluto entered Capricorn (cold sign), Uranus entered Aries (hot sign), and Neptune entered Pisces (cool sign).

Uranus square Pluto represents both extreme cold and extreme heat. Neptune in Pisces is extreme wetness. All three planets are associated with extreme weather: extratropical cyclones, hurricanes, tornadoes, heat waves, cold waves. Proponents of climate change relate rising greenhouse gases to increased intensity and frequency of hurricanes; planetary alignments reflect this, but not necessarily because of greenhouse gases. The same applies to extreme cold and heat. Watch the outer planets; when they form aspects to each other during a succession of years, weather becomes a hot topic because of many major storms and other weather extremes. When these aspects are few, weather is more or less "normal," with little that classifies as extreme.

On the other hand, this is not to say that humans are friends of the environment. They're definitely not. But humans, like the climate, are slow to change, and reluctant to pay a little more to own a hybrid or electric vehicle, install solar panels on their homes, take the bus to work, or reduce the consumption of hundreds of thousands of everyday items created from petroleum. Everyone needs to take a role in reducing greenhouse gases.

Conclusion

Success in weather forecasting requires practice and a basic knowledge of meteorology and local climatology. This can be achieved with time and effort. One of the best ways to begin is to choose a city, your own or another, and start writing forecasts.

Begin by learning the climatology of the location, and then work with the ingress and lunar phase charts. Check your results on the Weather Underground site (www.wunderground.com), where you can also research climatology and read the blogs to gain meteorology knowledge as it relates to the current weather. On the same site you can find a scientific forecaster discussion for most locations, which you can then correlate to the charts and aspects. These discussions are posted at least daily and sometimes more often, depending on current weather conditions.

Remember, the planets do not cause the weather; they reflect weather conditions and phenomena. Also remember that one aspect does not equate to a weather event and that a weather event is unlikely unless the potential is in an ingress or a lunar phase chart. Look for three or more factors in these charts and then watch the transiting Moon, which will often trigger the event. For example, rain showers could begin as the transiting Moon aspects the meridian or horizon of a chart that shows strong potential for rain because Venus in the chart aspects one of these angles.

Keep good notes of both your successes and challenges. As you gain knowledge, you can review these notes and determine why your forecast was off and probably find other factors that will help you refine future forecasts. The bonus: you'll be doing astrometeorology research and along the way will undoubtedly develop new techniques.

The information in this book covers the method I use, but there is still much to learn. Through your own study and practice, you will surely find other viable techniques to advance the collective knowledge.

Finally, don't be surprised if you at least somewhat revise your definition of "good" and "bad" weather. A cloudy day, a rain shower, or above- or below-average temperatures might flip over into the good category, just because your forecast was 100 percent accurate.

Most of all, have fun and enjoy the lovely Saturn days!

Glossary

declination: The angular distance north or south of the celestial equator.

horizon: The latitude of a location; the horizontal axis; the east/west axis traditionally referred to as the Ascendant/Descendant in a chart.

ingress: One of the four annual ingresses of the Sun into a cardinal sign that represent the beginning of a season; spring begins when the Sun enters Aries, summer begins when the Sun enters Cancer, autumn begins when the Sun enters Libra, and winter begins when the Sun enters Capricorn.

ingress chart: Used to forecast seasonal trends.

latitude: The degrees north or south of the equator; 1 degree of latitude equals about 1 degree of a sign, or about 50 miles.

longitude: The degrees west or east of Greenwich; 1 degree of longitude equals about 1 degree of a sign, or about 50 miles.

lunar phase: One of the four monthly phases of the Moon (new Moon, full Moon, and two quarter Moons).

lunar phase chart: Used to forecast for weekly and daily forecasts.

meridian: The longitude of a location; the vertical axis; the north/south axis traditionally referred to as the Midheaven/IC in a chart.

Resources

Weather Underground, www.wunderground.com. This extensive and outstanding website for weather also has many articles and blogs about meteorology, along with historical weather records, links to current meteorology data, and radar and satellite images.

WeatherGeek Pro, www.weathergeekpro.com. This mobile application provides easy access to meteorology sites.

Penn State e-Wall, www.meteo.psu.edu/~fxg1/ewall.html. This electronic map wall from the Department of Meteorology at Pennsylvania State University provides current data from numerous meteorology models used in forecasting.

National Hurricane Center, www.nhc.noaa.gov. Source for hurricane data and information.

Weather Prediction Center, www.hpc.ncep.noaa.gov. Provides current and archival data, and meteorology information.

Storm Prediction Center, www.spc.noaa.gov. Source for current storm and convective outlooks.

National Centers for Environmental Prediction, http://mag.ncep.noaa.gov. Source for meteorology tools.

Climate Prediction Center, www.cpc.ncep.noaa.gov/products/forecasts. Issues seasonal, long-range weather forecasts.

Grenci, Lee M., and John M. Nese. *A World of Weather: Fundamentals of Meteorology.* 5th ed. Dubuque, IA: Kendall/Hunt Pub. Co., 2012. This is an excellent book from which to learn the basics of meteorology.